Fuzzy Natural Logic in
IFSA-EUSFLAT 2021

Fuzzy Natural Logic in IFSA-EUSFLAT 2021

Editors

Antonin Dvorak
Vilém Novák

MDPI • Basel • Beijing • Wuhan • Barcelona • Belgrade • Manchester • Tokyo • Cluj • Tianjin

Editors
Antonin Dvorak
Institute for Research and
Applications of Fuzzy Modeling,
University of Ostrava,
Ostrava, Czech Republic

Vilém Novák
Institute for Research and
Applications of Fuzzy Modeling,
University of Ostrava,
Ostrava, Czech Republic

Editorial Office
MDPI
St. Alban-Anlage 66
4052 Basel, Switzerland

This is a reprint of articles from the Special Issue published online in the open access journal *Mathematics* (ISSN 2227-7390) (available at: https://www.mdpi.com/journal/mathematics/special_issues/IFSA_EUSFLAT2021).

For citation purposes, cite each article independently as indicated on the article page online and as indicated below:

LastName, A.A.; LastName, B.B.; LastName, C.C. Article Title. *Journal Name* **Year**, *Volume Number*, Page Range.

ISBN 978-3-0365-6147-9 (Hbk)
ISBN 978-3-0365-6148-6 (PDF)

© 2022 by the authors. Articles in this book are Open Access and distributed under the Creative Commons Attribution (CC BY) license, which allows users to download, copy and build upon published articles, as long as the author and publisher are properly credited, which ensures maximum dissemination and a wider impact of our publications.

The book as a whole is distributed by MDPI under the terms and conditions of the Creative Commons license CC BY-NC-ND.

Contents

About the Editors vii

Vilém Novák and Antonín Dvořák
Preface to the Special Issue on "Fuzzy Natural Logic in IFSA-EUSFLAT 2021"
Reprinted from: *Mathematics* **2022**, *10*, 4393, doi:10.3390/math10224393 1

Karel Fiala and Petra Murinová
A Formal Analysis of Generalized Peterson's Syllogisms Related to Graded Peterson's Cube
Reprinted from: *Mathematics* **2022**, *10*, 906, doi:10.3390/math10060906 5

Adrià Torrens-Urrutia, M. Dolores Jiménez-López, Antoni Brosa-Rodríguez, David Adamczyk
A Fuzzy Grammar for Evaluating Universality and Complexity in Natural Language
Reprinted from: *Mathematics* **2022**, *10*, 2602, doi:10.3390/math10152602 33

Adrià Torrens-Urrutia, Vilém Novák, María Dolores Jiménez-López
Describing Linguistic Vagueness of Evaluative Expressions Using Fuzzy Natural Logic and Linguistic Constraints
Reprinted from: *Mathematics* **2022**, *10*, 2760, doi:10.3390/math10152760 57

Jiří Janeček and Irina Perfilieva
Preimage Problem Inspired by the F-Transform
Reprinted from: *Mathematics* **2022**, *10*, 3209, doi:10.3390/math10173209 81

Paulo Vitor de Campos Souza, Edwin Lughofer, Huoston Rodrigues Batista, Augusto Junio Guimaraes
An Evolving Fuzzy Neural Network Based on Or-Type Logic Neurons for Identifying and Extracting Knowledge in Auction Fraud [†]
Reprinted from: *Mathematics* **2022**, *10*, 3872, doi:10.3390/math10203872 107

About the Editors

Antonin Dvorak

Antonín Dvořák is a senior researcher and associate professor at the Institute of Research and Applications of Fuzzy Modeling (University of Ostrava). He obtained his Ph.D. in Applied Mathematics at the University of Ostrava in 2004. He is the author and co-author of more than 60 scientific papers. He co-authored the following book by V. Novák, I. Perfilieva, and A. Dvořák: Insight into Fuzzy Modeling, published by John Wiley & Sons (2016). His research interests include mathematical fuzzy logic, the theory of aggreagation operators, approximate reasoning, and the theory of generalized quantifiers.

Vilém Novák

Vilém Novák is a senior researcher and full professor at the Institute of Research and Applications of Fuzzy Modeling (University of Ostrava). He is the founder and former director of this institute. His research interests include mathematical fuzzy logic and its applications. He is the author of several formal theories that served as source material for establishing the theory of fuzzy natural logic. His theoretical results led to the development of an original method of fuzzy/linguistic control of technological processes based on expert knowledge, which also has practical applications. He also developed several non-statistical methods for the processing of time series. He is the author and co-author of more than 300 scientific papers and 5 scientific monographs. His h-index is 29 (WOS).

Editorial

Preface to the Special Issue on "Fuzzy Natural Logic in IFSA-EUSFLAT 2021"

Vilém Novák * and Antonín Dvořák

Institute for Research and Applications of Fuzzy Modeling, University of Ostrava, 30. Dubna 22, 701 03 Ostrava, Czech Republic
* Correspondence: Vilem.Novak@osu.cz

In [1], we gave reasons to focus on one of the possible future directions of fuzzy logic towards the concept of fuzzy natural logic (FNL). The concept of FNL continues the program of fuzzy logic in a broader sense (FLb-logic) introduced in [2]. This theory was established as a formal logic aiming at modeling natural human reasoning, which necessarily proceeds in natural language. The goal is to make FNL a mathematical theory being extension of the mathematical fuzzy logic. Its paradigm extends the classical concept of natural logic suggested by Lakoff in [3]. According to him, natural logic is a collection of terms and rules that come with natural language and that allows us to reason and argue in it. Its goals can be characterized as follows:

- To express all concepts capable of being expressed in natural language,
- To characterize all the valid inferences that can be made in natural language,
- To mesh with adequate linguistic descriptions of all natural languages.

It is essential to employ meaning-postulates that do not vary from language to language. In other words, all natural languages reflect ability of human mind to reason that is common to all of us, and thus, its principles are independent on the use of the concrete natural language. The concept of natural logic has been further developed by several other authors (cf. [4,5] and others).

We argue that fuzzy set theory has the potential to be a good tool for modeling of linguistic semantics because it provides a reasonable mathematical model of the vagueness phenomenon. This is important because, as argued by many authors (cf., e.g., [6]), vagueness is an unavoidable feature of natural language semantics. The role of fuzzy sets in modeling of linguistic semantics has been discussed already by L. A. Zadeh in many of his papers since the very beginning (cf., e.g., [7–9]). Interesting is his concept of precisiated natural language [10]. Its main idea is to develop a "reasonable working formalization of the semantics of natural language without pretensions to capture it in detail and fineness." The goal is to provide an acceptable and applicable technical solution, i.e., to relax some of the requirements of thorough linguistic analysis and, in line with the paradigm of Zadeh's precisiated natural language, to focus on smaller parts of natural language and try to capture only their essential properties.

Therefore, following the definition of natural logic, we can define fuzzy natural logic as a system of theories of mathematical fuzzy logic enabling us to model terms and rules that come with natural language together with their inherent vagueness and allowing us to reason and argue using tools developed in it. A necessary constituent of FNL is a mathematical model of semantics of a specific part of natural language independent of a concrete language.

The following are the main sources for the development of FNL:

- Results of classical linguistics.
- Logical analysis of concepts and semantics of natural language Transparent Intensional Logic (P. Tichý [11], P. Materna [12]).
- Montague grammar [13].

Citation: Novák, V.; Dvořák, A. Preface to the Special Issue on "Fuzzy Natural Logic in IFSA-EUSFLAT 2021". *Mathematics* **2022**, *10*, 4393. https://doi.org/10.3390/math10224393

Received: 15 November 2022
Accepted: 16 November 2022
Published: 21 November 2022

Publisher's Note: MDPI stays neutral with regard to jurisdictional claims in published maps and institutional affiliations.

Copyright: © 2022 by the authors. Licensee MDPI, Basel, Switzerland. This article is an open access article distributed under the terms and conditions of the Creative Commons Attribution (CC BY) license (https://creativecommons.org/licenses/by/4.0/).

- Mathematical fuzzy logic, especially the higher order one called *Fuzzy Type Theory* (FTT) [14].

The current constituents of FNL are:

- Theory of evaluative linguistic expressions (small, very small, medium, large, etc.).
- Theory of fuzzy and intermediate quantifiers (most, a lot of, few, many, etc.) and generalized Aristotle's syllogisms.
- Theory of fuzzy/linguistic IF-THEN rules and logical inference (Perception-based Logical Deduction).

FNL is expected to contribute to the development of methods for construction of models of systems and processes on the basis of expert knowledge expressed in genuine natural language and to develop special algorithms making computer to "understand" natural language and suggest a corresponding behavior. Let us remark that FNL has already many interesting applications (cf. [15]).

There are five contributions in this Special Issue devoted to extended versions of the papers presented in the conference "The 19th World Congress of the International Fuzzy Systems Association and the 12th Conference of the European Society for Fuzzy Logic and Technology jointly with the AGOP, IJCRS, and FQAS conferences" that took place in Bratislava (Slovakia) from September 19 to September 24, 2021. These contributions use various parts and concepts of FNL mentioned above and apply it to a wide range of problems, theoretical as well as application-oriented.

A very important building block of FNL is the theory of evaluative linguistic expressions. In [16], this theory is developed in an exciting direction from the perspective of theoretical linguistics. The range of evaluative linguistic expressions is considerably broadened to also contain verbs ("love" in "I love you very much"), proper names ("Einstein" in "Mark is an Einstein"), etc. Essential for this extension is the Fuzzy Property Grammar—the topic of [17]. In this contribution, the Fuzzy Property Grammar permits to describe linguistic complexity of a natural language and linguistic universality (presence of a grammatical characteristics in all or most natural languages) as vague concepts. It allows, among other things, to better understand similarities and differences between natural languages.

One of the important directions of FNL development is the study of intermediate quantifiers and generalized syllogisms. In [18], the authors continue this research program by studying syllogisms whose constituents (quantified expressions) can contain negated terms, such as "most people who do not drink alcohol have healthy livers." The validity of certain forms of these syllogisms (related to the so-called graded Peterson's cube of opposition) is proved syntactically. Examples of syllogisms on finite models are also elaborated.

Paper [19] presents a more applied facet of FNL. It proposes a model for exploring and extracting knowledge of auction frauds using IF-THEN rules (a crucial component of FNL). An innovative fuzzy neural network model based on or-neurons using a t-conorm as their underlying operation is presented in detail and compared with several state-of-art neuro-fuzzy models. The proposed model shows its superiority by achieving more than 98% accuracy with fewer fuzzy rules and greater assertiveness than other models.

In [20], the authors study the so-called preimage problem in the context of F-transform: how it is possible to describe the class of all functions mapped onto the same result of direct F-transform. Note that F-transform is an important technique necessary in various kinds of applications of FNL. The relationship between objects is determined by closeness (a weaker concept than metric). The preimage problem is formulated using the language of matrix calculus. The authors show that its solutions can be given in three different ways (using a weighted arithmetic mean, any right inverse of the closeness matrix or any element of a certain affine subspace). The study of this problem contributes to better understanding of ill-posed problems frequent in machine learning.

We would like to cordially thank all authors, editors, anonymous referees and staff of MDPI for contributing to the creation of this Special Issue.

Funding: This research received no external funding.

Conflicts of Interest: The authors declare no conflict of interest.

References

1. Novák, V. Fuzzy Natural Logic: Towards Mathematical Logic of Human Reasoning. In *Fuzzy Logic: Towards the Future*; Seising, R., Trillas, E., Kacprzyk, J., Eds.; Springer: Berlin/Heidelberg, Germany, 2015; pp. 137–165.
2. Novák, V. Towards Formalized Integrated Theory of Fuzzy Logic. In *Fuzzy Logic and Its Applications to Engineering, Information Sciences, and Intelligent Systems*; Bien, Z., Min, K., Eds.; Kluwer: Dordrecht, The Netherlands, 1995; pp. 353–363.
3. Lakoff, G. Linguistics and natural logic. *Synthese* **1970**, *22*, 151–271. [CrossRef]
4. MacCartney, B.; Manning, C.D. An extended model of natural logic. In *IWCS-8'09 Proceedings of the Eight International Conference on Computational Semantics*; Association for Computational Linguistics: Stroudsburg, PA, USA, 2009; pp. 140–156.
5. van Benthem, J. A Brief History of Natural Logic. In *Logic, Navya-Nyaya and Applications, Homage to Bimal Krishna Matilal*; Chakraborty, M., Löwe, B., Nath Mitra, M., Sarukkai, S., Eds.; College Publications: London, UK, 2008.
6. van Rooij, R. Vagueness and Linguistics. In *Vagueness: A Guide*; Ronzitti, G., Ed.; Springer: Berlin/Heidelberg, Germany, 2011; pp. 123–170.
7. Zadeh, L.A. Quantitative Fuzzy Semantics. *Inf. Sci.* **1973**, *3*, 159–176. [CrossRef]
8. Zadeh, L.A. The concept of a linguistic variable and its application to approximate reasoning I, II, III. *Inf. Sci.* **1975**, *8–9*, 43–80; 199–257; 301–357. [CrossRef]
9. Zadeh, L.A. A computational approach to fuzzy quantifiers in natural languages. *Comput. Math. Appl.* **1983**, *9*, 149–184. [CrossRef]
10. Zadeh, L.A. Precisiated natural language. *AI Mag.* **2004**, *25*, 74–91.
11. Tichý, P. *The Foundations of Frege's Logic*; De Gruyter: Berlin, Germany, 1988.
12. Duží, M.; Jespersen, B.; Materna, P. *Procedural Semantics for Hyperintensional Logic*; Springer: Dordrecht, The Netherlands, 2010.
13. Dowty, D. *Word Meaning and Montague Grammar: The Semantics of Verbs and Times in Generative Semantics and in Montague's PTQ*; Springer: Berlin, Germany, 1979.
14. Novák, V. On Fuzzy Type Theory. *Fuzzy Sets Syst.* **2005**, *149*, 235–273. [CrossRef]
15. Novák, V.; Perfilieva, I.; Dvořák, A. *Insight into Fuzzy Modeling*; Wiley & Sons: Hoboken, NJ, USA, 2016.
16. Torrens-Urrutia, A.; Novák, V.; Jiménez-López, M.D. Describing Linguistic Vagueness of Evaluative Expressions Using Fuzzy Natural Logic and Linguistic Constraints. *Mathematics* **2022**, *10*, 2760. [CrossRef]
17. Torrens-Urrutia, A.; Jiménez-López, M.D.; Brosa-Rodríguez, A.; Adamczyk, D. A Fuzzy Grammar for Evaluating Universality and Complexity in Natural Language. *Mathematics* **2022**, *10*, 2602. [CrossRef]
18. Fiala, K.; Murinová, P. A Formal Analysis of Generalized Peterson's Syllogisms Related to Graded Peterson's Cube. *Mathematics* **2022**, *10*, 906. [CrossRef]
19. de Campos Souza, P.V.; Lughofer, E.; Batista, H.R.; Guimaraes, A.J. An Evolving Fuzzy Neural Network Based on Or-Type Logic Neurons for Identifying and Extracting Knowledge in Auction Fraud. *Mathematics* **2022**, *10*, 3872. doi: 10.3390/math10203872. [CrossRef]
20. Janeček, J.; Perfilieva, I. Preimage Problem Inspired by the F-Transform. *Mathematics* **2022**, *10*, 3209. [CrossRef]

Article

A Formal Analysis of Generalized Peterson's Syllogisms Related to Graded Peterson's Cube

Karel Fiala [1,†] and Petra Murinová [2,*,†]

1. Department of Mathematics, University of Ostrava, 702 00 Ostrava, Czech Republic; karel.fiala@osu.cz
2. Institute for Research and Application of Fuzzy Modeling, University of Ostrava, 702 00 Ostrava, Czech Republic
* Correspondence: petra.murinova@osu.cz
† These authors contributed equally to this work.

Abstract: This publication builds on previous publications in which we constructed syntactic proofs of fuzzy Peterson's syllogisms related to the graded square of opposition. The aim of the publication is to be formally able to find syntactic proofs of fuzzy Peterson's logical syllogisms with forms of fuzzy intermediate quantifiers that design the graded Peterson's cube of opposition.

Keywords: fuzzy Peterson's syllogisms; fuzzy intermediate quantifiers; graded Peterson's cube of opposition

MSC: 03B16, 03B38, 03B50

Citation: Fiala, K.; Murinová, P. A Formal Analysis of Generalized Peterson's Syllogisms Related to Graded Peterson's Cube. *Mathematics* 2022, 10, 906. https://doi.org/10.3390/math10060906

Academic Editor: Manuel Ojeda-Aciego

Received: 27 January 2022
Accepted: 7 March 2022
Published: 11 March 2022

Publisher's Note: MDPI stays neutral with regard to jurisdictional claims in published maps and institutional affiliations.

Copyright: © 2022 by the authors. Licensee MDPI, Basel, Switzerland. This article is an open access article distributed under the terms and conditions of the Creative Commons Attribution (CC BY) license (https://creativecommons.org/licenses/by/4.0/).

1. Introduction

The article aims to follow up on the achieved results concerning the formal proofs of fuzzy Peterson syllogisms.

The theory of syllogistic reasoning was investigated by several authors as a generalization of classical Aristotelian syllogisms ([1–3]). Categorical syllogisms ([4,5]) consist of three main parts: *the major premise*, *the minor premise*, and *the conclusion*. With the introduction of the term "generalized quantifier", which was proposed by Mostowski in [6], the study of logical syllogisms expanded to a group of generalized logical syllogisms. Such syllogisms include forms that contain different types of generalized quantifiers. One special group is intermediate quantifiers. This topic started to be interesting for linguists and philosophers, who laid a question of how to explain expressions that represent generalized quantifiers.

Peterson, in his book [7], is interested in the group of intermediate quantifiers which lie between classical quantifiers. He first philosophically analyzed and explained the meaning of intermediate quantifiers in terms of their position in the generalized square of opposition. Furthermore, he continued the study of the group of generalized logical syllogisms. Peterson's enlargement was established on an idea to substitute a classical quantifier with an intermediate one in the classical four figures, which returned in 105 new valid intermediate syllogisms. A check of the validity of a group of logical syllogisms was conducted using Venn diagrams, which was carried out by several authors [7–9]. Below, we present an example of a non-trivial fuzzy intermediate syllogism as follows:

P_1 : *Almost all people do not have a plane.*
P_2 : *Most people have a phone.*
C : *Some people who have a phone do not have a plane.*

The work of authors who dealt with generalized quantifiers was followed by several authors with the advent of the definition of a fuzzy set. Several authors followed up this approach. They introduced several forms of logical syllogisms with *fuzzy* generalized quantifiers. In 1985, L. Zadeh semantically interpreted a special group of fuzzy syllogisms with fuzzy intermediate quantifiers in both premises as well as in the conclusion. Below, we present a very famous *multiplicative chaining syllogism* ([10]) as follows:

$$\frac{Q_1 \ Y \text{ is } M}{Q_2 \ M \text{ is } X}$$
$$\frac{Q_2 \ M \text{ is } X}{(\geq Q_1 \otimes Q_2) \ Y \text{ are } X.}$$

As was explained by Zadeh, the expression $\geq Q_1 \otimes Q_2$ in the conclusion of Zadeh's syllogism can be read as "at least $Q_1 \otimes Q_2$".

Zadeh's work was later extended and sophisticated by many authors. From the point of view of fuzzy quantifiers, which are represented as intervals in Didier Duboise's approach in ([11,12]), Zadeh's special syllogism is compared by M. Pereira-Fariña in [13]. In the publications, Dubois et al. work with quantifiers represented as crisp closed intervals (*more than a half* $= [0.5, 1]$, *around five* $= [4, 6]$). A typical example of such interval fuzzy syllogism appears as follows:

$$\frac{P_1 \colon [5\%, 10\%] \text{ students have a job.}}{P_2 \colon [5\%, 10\%] \text{ students have a child.}}$$
$$\frac{P_2 \colon [5\%, 10\%] \text{ students have a child.}}{C \colon [0\%, 10\%] \text{ students have a child and have a job.}}$$

Recall that a global overview of fuzzy generalized quantifiers and the various mechanisms for defining these quantifiers can be found in [14].

M. Pereira-Fariña et al. follow, in the next publication [15], by interpreting logical syllogisms with more premises. In this publication, a group of authors suggested a *general inference schema for syllogistic reasoning*, which was proposed as the transformation of the syllogistic reasoning study into an equivalent optimization problem.

The above-mentioned publications study and verify the validity of fuzzy syllogisms, especially semantically. In [16], a mathematical definition of fuzzy intermediate quantifiers based on the theory of evaluative linguistic expressions was proposed. The motivation, fundamental assumptions, and formalization of this theory are described in detail in [17]. Later, in [18], we focused on the syntactic construction of proofs of all 105 basic fuzzy logical syllogisms that relate to the graded Peterson's square of opposition (see [19]). Typical examples of natural language expressions contained in Peterson's logical syllogisms are as follows:

Most children like computer games.
Most cats like to sleep.

1.1. Application of Generalized Quantifiers

Generalized quantifiers offer several types of applications in economics, medicine, heavy industry, biology, etc. Let us first mention applications related to a group of fuzzy intermediate quantifiers which are represented by natural language expressions. One of the interesting applications is time series prediction, which has its application mainly in economic fields. In [20], the author proposed an interpretation, forecasting, and linguistic characterization of time series. The result makes it possible to obtain information about the data using natural language, which is much more understandable to the average user. To illustrate the reader, we present an example of the linguistic interpretation of economic time series, using natural language, as follows:

Most (many, few) analyzed time series stagnated recently, but their future trend is slightly increasing [20].

Another area that is very closely related to this application is getting new information from natural language data. Here, it is offered to use the theory of syllogistic reasoning and to obtain new information from natural language data using valid forms of syllogisms. Time series were also analyzed by a group of authors in [21]. There are also publications of authors who are interested in the linguistic summarization of data.

In [22], the authors introduced methods for using the linguistic database to summarize natural data (see [23–26]). Later, these methods were improved in [27] and implemented by Kasprzyk and Zadrożny [28]. Another linguistic summarization of process data was proposed in [29].

Another area of application is the use of fuzzy association analysis and the use of association rules to interpret natural language data. An algorithm for the interpretation of biological data using fuzzy intermediate quantifiers was proposed in [30].

Most irises with both small-length sepals and petals have small-width petals.

1.2. Main Goals

The main idea of this publication is to work with terms that also contain negated terms in the antecedent, and to study related valid fuzzy syllogisms. Typical examples of natural language expressions which are related to the graded Peterson's ([19]) cube are as follows:

Almost all students who do not like mathematics do not study technical fields.
Most people who do not drink alcohol have healthy livers.

A typical example of a syllogism with fuzzy intermediate quantifiers in both premises, which is called *non-trivial*, reads as follows:

P_1 : *Almost all people who do sports have healthy lungs.*
P_2 : *Almost all people who do sports do not have asthma.*
C : *Some people who do not have asthma have healthy lungs.*

1.3. Application of New Forms of Fuzzy Intermediate Quantifiers

As mentioned above, there are several application areas where natural language expressions are used. We also gave, in the previous section, specific examples of natural language expressions that occur in both of the premises of syllogisms or are used to interpret natural data. Therefore, the idea is offered to first find and formally prove the validity of new forms of logical syllogisms, and further, to work on the use of these forms in the areas of fuzzy association analysis, language interpretation, linguistic summarization, the interpretation of time series, etc.

The paper is structured as follows: after the motivational introduction, the reader is acquainted with mathematical theory in the methods section. The third section contains important mathematical definitions of new forms of intermediate quantifiers. In this section, we follow with proofs of new forms of logical syllogisms. The results section is closed by concrete examples of valid forms of logical syllogisms. This section continues with a discussion section, in which we summarize the results achieved. We conclude this paper with a conclusion and statement of future directions.

2. Main Methods

The main approach of this section is to recall the theory of fuzzy natural logic (FNL) that was designed based on the fuzzy type theory (FTT). FNL is a formal mathematical theory that includes three theories:

- Theory of evaluative linguistic expressions (see [17]);
- Theory of fuzzy IF–THEN rules and approximate reasoning (see [31]);
- Theory of intermediate quantifiers, generalized syllogisms, and graded structures of opposition (see [16,18]).

2.1. Fuzzy Type Theory

This section focuses on the reminder of the main symbols and of the fuzzy type theory. We will not repeat all the details here; we refer readers to previous publications [17,32].

Let us recall, at this point, that the mathematical theory of fuzzy quantifiers was proposed over the Łukasiewicz fuzzy type theory (Ł-FTT). The structure of truth values is represented by a linearly ordered MV_Δ-algebra that is extended by the delta operation (see [33,34]). A particular case is the standard Łukasiewicz MV_Δ-algebra:

$$\mathcal{L} = \langle [0,1], \vee, \wedge, \otimes, \rightarrow, 0, 1, \Delta \rangle. \tag{1}$$

The fundamental objects that represent the syntax of Ł-FTT are classical (cf. [35]). We assume atomic types as follows: ϵ (elements) and o (truth values). General types are marked by Greek letters $\alpha, \beta, \ldots$. A set of all types is marked by *Types*. The (meta-)symbol ":=" used below means "is defined by".

The *language* contains of variables $x_\alpha, \ldots$, special constants $c_\alpha, \ldots$ ($\alpha \in \textit{Types}$), symbol λ, and parentheses. The connectives are *fuzzy equality/equivalence* $\equiv$, *conjunction* $\wedge$, *implication* $\Rightarrow$, *negation* $\neg$, *strong conjunction* &, *strong disjunction* $\triangledown$, *disjunction* $\vee$, and *delta* Δ. The fuzzy type theory is *complete*, i.e., a theory T is consistent iff it has a (Henkin) model ($\mathcal{M} \models T$). We sometimes use the equivalent notion: $T \vdash A_o$ iff $T \models A_o$.

The following special formulas are important in our theory below:

$$\Upsilon_{oo} \equiv \lambda z_o \cdot \neg \Delta(\neg z_o), \qquad \text{(nonzero truth value)}$$
$$\hat{\Upsilon}_{oo} \equiv \lambda z_o \cdot \neg \Delta(z_o \vee \neg z_o). \qquad \text{(general truth value)}$$

Thus, $\mathcal{M}(\Upsilon(A_o)) = 1$ iff $\mathcal{M}(A_o) > 0$, and $\mathcal{M}(\hat{\Upsilon}(A_o)) = 1$ iff $\mathcal{M}(A_o) \in (0,1)$ holds in any model $\mathcal{M}$.

2.2. Evaluative Linguistic Expressions

As we stated in the introduction, the formal definitions of fuzzy intermediate quantifiers are based on evaluation linguistic expressions. In this subsection, we recall the theory of evaluative linguistic expressions.

Evaluative linguistic expressions are special expressions of a natural language, for example, *very small, roughly medium, extremely large, very long, quite roughly, extremely rich*, etc. Their theory is the main part of the fuzzy natural logic. The evaluative language expressions themselves play an important role in everyday human reasoning. We can find them in a wide variety of areas, such as economics, decision making, and more.

The language J^{Ev} contains special symbols as follows:

- The constants $\top, \bot \in \textit{Form}_o$ for truth and falsity, and $\dagger \in \textit{Form}_o$ for the middle truth value;
- A special constant $\sim \in \textit{Form}_{(oo)o}$ for an additional fuzzy equality on the set of truth values L;
- A set of special constants $\nu, \ldots \in \textit{Form}_{oo}$ for linguistic hedges. The J^{Ev} contains the following special constants: $\{Ex, Si, Ve, ML, Ro, QR, VR\}$' these denote the linguistic hedges: (*extremely, significantly, very, roughly, more or less, rather, quite roughly*, and *very roughly*, respectively);
- A set of triples of next constants $(\mathbf{a}_\nu, \mathbf{b}_\nu, \mathbf{c}_\nu), \ldots \in \textit{Form}_o$, where each hedge ν is uniquely connected with one triple of these constants.

The evaluative expressions are interpreted by special formulas $Sm \in \textit{Form}_{oo(oo)}$ (small), $Me \in \textit{Form}_{oo(oo)}$ (medium), $Bi \in \textit{Form}_{oo(oo)}$ (big), and $Ze \in \textit{Form}_{oo(oo)}$ (zero) that can be expanded by several linguistic hedges. Let us remind that a *hedge*, which is often an adverb such as "extremely, significantly, very, roughly", etc., is, in general, represented by a formula $\nu \in \textit{Form}_{oo}$ with specific properties. We proposed a formula $Hedge \in \textit{Form}_{o(oo)}$. Then, $T^{Ev} \vdash Hedge\, \nu$ means that ν is a hedge. The other details can be found in [17]. The formula $T^{Ev} \vdash Hedge\, \nu$ is provable for all $\nu \in \{Ex, Si, Ve, ML, Ro, QR, VR\}$. Furthermore, evaluative linguistic expressions are represented by formulas:

$$Sm\,\nu, Me\,\nu, Bi\,\nu, Ze\,\nu \in \textit{Form}_{oo}, \tag{2}$$

where ν is a hedge. We will also assume an *empty hedge* $\bar{\nu}$ that is introduced in front of *small*, *medium*, and *big* if no other hedge is assumed. A special hedge is Δ_{oo}, that represents the expression "utmost" and occurs below in the evaluative expression $Bi\,\Delta$.

Let $\nu_{1,oo}, \nu_{2,oo}$ be two hedges, i.e., $T^{Ev} \vdash Hedge\ \nu_{1,oo}$ and $T^{Ev} \vdash Hedge\ \nu_{2,oo}$. We propose a relation of the partial ordering of hedges by:

$$\ll := \lambda p_{oo} \lambda q_{oo} \cdot (\forall z_o)(p_{oo}z \Rightarrow q_{oo}z).$$

Lemma 1 ([17]). *The following ordering of the specific hedges can be proved.*

$$T^{Ev} \vdash \Delta \ll Ex \ll Si \ll Ve \ll \bar{\nu} \ll ML \ll Ro \ll QR \ll VR. \tag{3}$$

Theorem 1. *If* $T^{IQ} \vdash \nu_1 \ll \nu_2$, *then*

$$T^{IQ} \vdash (Bi\,\nu_1)((\mu(\neg B))(\neg B|z)) \Rightarrow (Bi\,\nu_2)((\mu(\neg B))(\neg B|z)).$$

Proof. Analogously to [16], this can be proved using Theorem 1(g), by replacing B with its negation. □

Evaluative expressions represent certain unspecified positions on a bounded linearly ordered scale. It is important to introduce the *context* in which we characterize them. The context can be characterized by a function $w : L \to N$ for some set N. We suggest the context as a triple of numbers $v_L, v_S, v_R \in N$, such that $v_L < v_S < v_R$ (the ordering on N is induced by w). Then, $x \in w$ iff $x \in [v_L, v_S] \cup [v_S, v_R]$. Introducing the concept of context means defining the concept of intension and extension. The intension of the evaluative linguistic expressions (2) is equal to a simple fuzzy set in the support of truth values. For further details, we recommend readers to the publication [17].

2.3. Fuzzy Measure

As discussed in our previous publications, the semantics of the intermediate quantifiers assumes the idea of a "size" of a (fuzzy) set, which we describe by the concept of a fuzzy measure. In our theory, we will assume the fuzzy measure below:

Definition 1. *Let* $R \in Form_{o(o\alpha)(o\alpha)}$ *be a formula.*

- *A formula,* $\mu \in Form_{o(o\alpha)(o\alpha)}$, *defined by:*

$$\mu_{o(o\alpha)(o\alpha)} := \lambda z_{o\alpha} \lambda x_{o\alpha} (R z_{o\alpha}) x_{o\alpha}, \tag{4}$$

represents a measure on fuzzy sets *in the universe of type* $\alpha \in Types$ *if it has the following properties:*
1. $\Delta(x_{o\alpha} \subseteq z_{o\alpha})\ \&\ \Delta(y_{o\alpha} \subseteq z_{o\alpha})\ \&\ \Delta(x_{o\alpha} \subseteq y_{o\alpha}) \Rightarrow ((\mu z_{o\alpha})x_{o\alpha} \Rightarrow (\mu z_{o\alpha})y_{o\alpha});$
2. $\Delta(x_{o\alpha} \subseteq z_{o\alpha}) \Rightarrow ((\mu z_{o\alpha})(z_{o\alpha} \setminus x_{o\alpha}) \equiv \neg(\mu z_{o\alpha})x_{o\alpha});$
3. $\Delta(x_{o\alpha} \subseteq y_{o\alpha})\ \&\ \Delta(x_{o\alpha} \subseteq z_{o\alpha})\ \&\ \Delta(y_{o\alpha} \subseteq z_{o\alpha}) \Rightarrow ((\mu z_{o\alpha})x_{o\alpha} \Rightarrow (\mu y_{o\alpha})x_{o\alpha}).$

The fuzzy measure introduced above is defined using three axioms—the axiom of normality, the axiom of monotonicity, and the fuzzy measure is closed with respect to the negation.

Example 1. *A fuzzy measure on a finite universe can be introduced as follows. Let M be a finite set and $A, B \subseteq M$ be fuzzy sets. Put:*

$$|A| = \sum_{m \in \mathrm{Supp}(A)} A(m). \tag{5}$$

Furthermore, let us define a function $F^R \in (L^{\mathcal{F}(M)})^{\mathcal{F}(M)}$ by:

$$F^R(B)(A) = \begin{cases} 1, & \text{if } B = A = \emptyset, \\ \min\left\{1, \frac{|A|}{|B|}\right\}, & \text{if } \mathrm{Supp}(A) \subseteq \mathrm{Supp}(B), \\ 0, & \text{otherwise} \end{cases} \quad (6)$$

for all $A, B \subseteq_\sim M$.

2.4. Formal Definition of Intermediate Quantifiers

In this subsection, we will recall the modified definition of the fuzzy intermediate quantifier, which is based on a special fuzzy set representing the cut of a fuzzy set (support).

In this article, we will work with the special fuzzy sets; they represent "cuts" of the universe B.

Let $y, z \in Form_{o\alpha}$. The cut of y by z is the fuzzy set:

$$y|z \equiv \lambda x_\alpha \cdot zx \,\&\, \Delta(Y(zx) \Rightarrow (yx \equiv zx)).$$

The following result can be proved.

Proposition 1. *Let $\mathcal{M}$ be a model such that $B = \mathcal{M}(y) \subseteq_\sim M_\alpha$, $Z = \mathcal{M}(z) \subseteq_\sim M_\alpha$. Then, for any $m \in M_\alpha$:*

$$\mathcal{M}(y|z)(m) = (B|Z)(m) = \begin{cases} B(m), & \text{if } B(m) = Z(m), \\ 0 & \text{otherwise}. \end{cases}$$

We can observe that the operation $B|Z$ picks only those elements $m \in M_\alpha$ from the support B, whose membership degree $B(m)$ is equal to the degree $Z(m)$; otherwise, it is equal to zero $((B|Z)(m) = 0)$. If there is no such element, then the cut is represented by an empty set ($B|Z = \emptyset$).

Definition 2. *Let Ev be a formula representing an evaluative expression, x be a variable and A, B, z be formulas. Then, either of the formulas:*

$$(Q_{Ev}^\forall x)(B, A) \equiv (\exists z)[(\forall x)((B|z)\, x \Rightarrow Ax) \wedge Ev((\mu B)(B|z))], \quad (7)$$

$$(Q_{Ev}^\exists x)(B, A) \equiv (\exists z)[(\exists x)((B|z)x \wedge Ax) \wedge Ev((\mu B)(B|z))], \quad (8)$$

construe the sentence:

"⟨Quantifier⟩ Bs are A".

Below, we introduce several examples of fuzzy intermediate quantifiers which fulfill the property of the monotonicity.

A: All Bs are $A := (Q^\forall_{Bi\Delta} x)(B, A) \equiv (\forall x)(Bx \Rightarrow Ax)$;

E: No Bs are $A := (Q^\forall_{Bi\Delta} x)(B, \neg A) \equiv (\forall x)(Bx \Rightarrow \neg Ax)$;

P: Almost all Bs are $A := (Q^\forall_{Bi\,Ex} x)(B, A)$;

B: Almost all Bs are not $A := (Q^\forall_{Bi\,Ex} x)(B, \neg A)$;

T: Most Bs are $A := (Q^\forall_{Bi\,Ve} x)(B, A)$;

D: Most Bs are not $A := (Q^\forall_{Bi\,Ve} x)(B, \neg A)$;

K: Many Bs are $A := (Q^\forall_{\neg Sm} x)(B, A)$;

G: Many Bs are not $A := (Q^\forall_{\neg Sm} x)(B, \neg A)$;

F: A few (A little) Bs are $A := (Q^\forall_{Sm\,Si} x)(B, A)$;

V: A few (A little) Bs are not $A := (Q^\forall_{Sm\,Si} x)(B, \neg A)$;

S: Several Bs are $A := (Q^\forall_{Sm\,Ve} x)(B, A)$;

Z: Several Bs are not $A := (Q^\forall_{Sm\,Ve} x)(B, \neg A)$;

I: Some Bs are $A := (Q^\exists_{Bi\Delta} x)(B, A) \equiv (\exists x)(Bx \wedge Ax)$;

O: Some Bs are not $A := (Q^\exists_{Bi\Delta} x)(B, \neg A) \equiv (\exists x)(Bx \wedge \neg Ax)$.

The mathematical definition of fuzzy intermediate quantifiers is extended by a formula that ensures the non-emptiness of the fuzzy set representing the antecedent.

Definition 3 ([19]). *Let Ev be a formula representing an evaluative expression, x be a variable, and A, B, z be formulas. Then, either of the formulas:*

$$(^*Q^\forall_{Ev} x)(B, A) \equiv (\exists z)[(\forall x)((B|z) x \Rightarrow Ax) \& (\exists x)(B|z)x \wedge Ev((\mu B)(B|z))], \quad (9)$$

$$(^*Q^\exists_{Ev} x)(B, A) \equiv (\exists z)[(\exists x)((B|z)x \wedge Ax) \triangledown \neg (\exists x)(B|z)x \wedge Ev((\mu B)(B|z))], \quad (10)$$

construe the sentence:

  "⟨*Quantifier⟩ Bs are A".

The corresponding quantifiers with presuppositions are denoted by *A, *E, *P, *B, *T, *D, *K, *G, *F, *V, *S, *Z, *I, and *O.

3. Results

3.1. Formal Structure of Peterson's Syllogisms Related to Peterson's Square

Definition 4 (Syllogism). *A syllogism is a triple $\langle P_1, P_2, C \rangle$ of three statements. P_1, P_2 are called premises (P_1 represents major, P_2 is minor) and C denotes a conclusion. S (subject) is somewhere in P_2 and also as the first formula of the conclusion C, formula P (predicate) is somewhere in P_1 and as the second formula of C; a formula that is not introduced in the conclusion C is called a middle formula M.*

Fuzzy syllogisms are obtained by replacing the classical quantifier (or classical quantifiers) with the fuzzy quantifier (or fuzzy quantifiers).

Definition 5. *Syllogism $\langle P_1, P_2, C \rangle$ is strongly valid in T^{IQ} if $T^{IQ} \vdash (P_1 \& P_2) \Rightarrow C$, or equivalently, if $T^{IQ} \vdash P_1 \Rightarrow (P_2 \Rightarrow C)$.*

Naturally speaking, the syllogism is valid if the Łukasiewicz conjunction of the degrees of both premises is less than or equal to the value of the conclusion.

Since we assume one middle formula, we can therefore consider four possible figures, which will arise according to the position of the middle formula.

Definition 6. *Let Q_1, Q_2, Q_3 be fuzzy quantifiers and $M, P, S \in Form_{o\alpha}$ be formulas. Let M be a middle formula, S be a subject, and P be a predicate. Then, we distinguish four corresponding figures:*

Figure I	*Figure II*	*Figure III*	*Figure IV*
Q_1 M are P	Q_1 P are M	Q_1 M are P	Q_1 P are M
Q_2 S are M	Q_2 S are M	Q_2 M are S	Q_2 M are S
Q_3 S are P	Q_3 S are P	Q_3 S are P	Q_3 S are P

For demonstration, we present an overview of the valid Peterson's logical syllogisms of Figure I, which were formally proved in [18]. We can observe that the main role plays the property of the monotonicity.

Theorem 2 ([18]). *The following syllogisms are strongly valid in T^{IQ}:*

AAA						
AAP	APP					
AAT	APT	ATT				
AAK	APK	ATK	AKK			
AAF	APF	ATF	AKF	AFF		
AAS	APS	ATS	AKS	AFS	ASS	
A(*A)I	A(*P)I	A(*T)I	A(*K)I	A(*F)I	A(*S)I	AII

Below, we present an example of valid logical syllogism **AKK**-I with the fuzzy intermediate quantifiers.

All cows are herbivores.
Many animals on the farm are cows.
Many animals on the farm are herbivores.

We follow with the negative syllogisms of Figure I.

Theorem 3 ([18]). *The following syllogisms are strongly valid in T^{IQ}:*

EAE						
EAB	EPB					
EAD	EPD	ETD				
EAG	EPG	ETG	EKG			
EAV	EPV	ETV	EKV	EFV		
EAZ	EPZ	ETZ	EKZ	EFZ	ESZ	
E(*A)O	E(*P)O	E(*T)O	E(*K)O	E(*F)O	E(*S)O	EIO

At this point, we will not present other figures and their valid syllogisms. We refer readers to publications in which we have addressed the formal construction of mathematical proofs of all of Peterson's syllogisms (see [18]).

3.2. New Forms of Fuzzy Intermediate Quantifiers Related to Graded Peterson's Cube

Below, we introduce the mathematical definitions of intermediate quantifiers, which form a graded Peterson's cube of opposition (see Figure A2).

Definition 7 (New forms of fuzzy intermediate quantifiers). *Let Ev be a formula representing an evaluative expression, x be a variable, and A, B, z be formulas. Then, for either of the formulas:*

$$(Q^{\forall}_{Ev}x)(\neg B, \neg A) \equiv (\exists z)[(\forall x)((\neg B|z)\,x \Rightarrow \neg Ax) \wedge Ev((\mu(\neg B))(\neg B|z))], \qquad (11)$$

$$(Q^{\exists}_{Ev}x)(\neg B, \neg A) \equiv (\exists z)[(\exists x)((\neg B|z)x \wedge \neg Ax) \wedge Ev((\mu(\neg B))(\neg B|z))], \qquad (12)$$

either of the quantifiers (11) *or* (12) *construes the sentence*
   "⟨*quantifier*⟩ *not Bs are not A*".

Below, we introduce the list of several forms of fuzzy intermediate quantifiers.

$$\textbf{a: All } \neg Bs \text{ are not } A := (Q_{Bi\Delta}^{\forall} x)(\neg B, \neg A) \equiv (\forall x)(\neg Bx \Rightarrow \neg Ax);$$

$$\textbf{e: No } \neg Bs \text{ are not } A := (Q_{Bi\Delta}^{\forall} x)(\neg B, A) \equiv (\forall x)(\neg Bx \Rightarrow Ax);$$

$$\textbf{p: Almost all } \neg Bs \text{ are not } A := (Q_{BiEx}^{\forall} x)(\neg B, \neg A);$$

$$\textbf{b: Almost all } \neg Bs \text{ are } A := (Q_{BiEx}^{\forall} x)(\neg B, A);$$

$$\textbf{t: Most } \neg Bs \text{ are not } A := (Q_{BiVe}^{\forall} x)(\neg B, \neg A);$$

$$\textbf{d: Most } \neg Bs \text{ are } A := (Q_{BiVe}^{\forall} x)(\neg B, A);$$

$$\textbf{k: Many } \neg Bs \text{ are not } A := (Q_{\neg Sm}^{\forall} x)(\neg B, \neg A);$$

$$\textbf{g: Many } \neg Bs \text{ are } A := (Q_{\neg Sm}^{\forall} x)(\neg B, A);$$

$$\textbf{f: A few (A little) } \neg Bs \text{ are not } A := (Q_{SmSi}^{\forall} x)(\neg B, \neg A);$$

$$\textbf{v: A few (A little) } \neg Bs \text{ are } A := (Q_{SmSi}^{\forall} x)(\neg B, A);$$

$$\textbf{s: Several } \neg Bs \text{ are not } A := (Q_{SmVe}^{\forall} x)(\neg B, \neg A);$$

$$\textbf{z: Several } \neg Bs \text{ are } A := (Q_{SmVe}^{\forall} x)(\neg B, A)$$

$$\textbf{i: Some } \neg Bs \text{ are not } A := (Q_{Bi\Delta}^{\exists} x)(\neg B, \neg A) \equiv (\exists x)(\neg Bx \wedge \neg Ax);$$

$$\textbf{o: Some } \neg Bs \text{ are } A := (Q_{Bi\Delta}^{\exists} x)(\neg B, A) \equiv (\exists x)(\neg Bx \wedge Ax).$$

If the presupposition is needed, we will denote the corresponding quantifiers by *a, *e, *p, *b, *t, *d, *k, *g, *f, *v, *s, *z, *i, and *o

Just as the theorem represents the monotonicity of the quantifiers that form Peterson's square of opposition, we can also prove the monotonic behavior for new forms of quantifiers.

Theorem 4. *The set of implications below is provable in Ł-FTT:*

1. $T^{IQ} \vdash a \Rightarrow p, T^{IQ} \vdash p \Rightarrow t, T^{IQ} \vdash t \Rightarrow k,$
   $T^{IQ} \vdash k \Rightarrow f, T^{IQ} \vdash f \Rightarrow s, T^{IQ} \vdash s \Rightarrow i;$
2. $T^{IQ} \vdash e \Rightarrow b, T^{IQ} \vdash b \Rightarrow d, T^{IQ} \vdash d \Rightarrow g,$
   $T^{IQ} \vdash g \Rightarrow v, T^{IQ} \vdash v \Rightarrow z, T^{IQ} \vdash z \Rightarrow o.$

Proof. We will show the proof of $T^{IQ} \vdash \textbf{p} \Rightarrow \textbf{t}$. We know that:

$$T^{IQ} \vdash (\forall x)((\neg B|z)\, x \Rightarrow \neg Ax) \Rightarrow (\forall x)((\neg B|z)\, x \Rightarrow \neg Ax). \tag{13}$$

By Lemma 1, we know that $Ex \ll Ve$ and, therefore, from Theorem 1 and from (13), we obtain:

$$T^{IQ} \vdash [(\forall x)((\neg B|z)\, x \Rightarrow \neg Ax) \wedge BiEx((\mu(\neg B))(\neg B|z))] \Rightarrow$$
$$\Rightarrow [(\forall x)((\neg B|z)\, x \Rightarrow \neg Ax) \wedge BiVe((\mu(\neg B))(\neg B|z))].$$

Then, by generalization ($\forall z$) and the properties of the quantifiers, we obtain

$$T^{IQ} \vdash (\exists z)[(\forall x)((\neg B|z)\, x \Rightarrow \neg Ax) \wedge BiEx((\mu(\neg B))(\neg B|z))] \Rightarrow$$
$$\Rightarrow (\exists z)[(\forall x)((\neg B|z)\, x \Rightarrow \neg Ax) \wedge BiVe((\mu(\neg B))(\neg B|z))].$$

This proof is analogous to the proof of $T^{IQ} \vdash \mathbf{P} \Rightarrow \mathbf{T}$ - we only replace each formula in the proof by its negation. The monotonicity of fuzzy intermediate quantifiers is ensured by the monotonicity of evaluative linguistic expressions. The other proofs of implications are similar to the proof of Theorem A2; we only replace each formula in the proof with its negation. □

There are valid forms of examples of logical syllogisms with new forms of fuzzy intermediate quantifiers related to a graded Peterson's cube of opposition (see Appendix A).

g: Many animals which are not mammals are fish.
A: All dolphins are mammals.
o: Some animals which are not dolphins are fish.

g: Many diseases which are not lethal are virus diseases.
E: All virus diseases can not be cured by antibiotics.
i: Some diseases which can not be cured by antibiotics are not lethal diseases.

3.3. Valid Forms Related to Second Face

First, the reader is reminded of the valid syllogisms of the first figure that are related to the second face of the graded Peterson's cube of opposition. It is not necessary to construct mathematical proofs, because the validity of syllogisms can be verified very easily by replacing individual formulas with their negations.

Theorem 5. *The following syllogisms are strongly valid in* T^{IQ}:

aaa					
aap	app				
aat	apt	att			
aak	apk	atk	akk		
aaf	apf	atf	akf	aff	
aas	aps	ats	aks	afs	ass
a(*a)i	a(*p)i	a(*t)i	a(*k)i	a(*f)i	a(*s)i	aii

Proof. Analogously to Theorem 2, this can be proved by replacing each formula with its negation. □

Theorem 6. *The following syllogisms are strongly valid in* T^{IQ}:

eae					
eab	epb				
ead	epd	etd			
eag	epg	etg	ekg		
eav	epv	etv	ekv	efv	
eaz	epz	etz	ekz	efz	esz
e(*a)o	e(*p)o	e(*t)o	e(*k)o	e(*f)o	e(*s)o	eio

Proof. Analogously to Theorem 3, this can be proved by replacing each formula by its negation. □

Similarly, we can verify the validity of other forms of logical syllogisms of the other figures. We will not repeat other figures at this point.

3.4. New Forms of Figure I

In the previous part of this paper, we showed strongly valid syllogisms of the first face and strongly valid syllogisms of the second face. Another goal of the publication is to verify the validity of logical syllogisms that describe the relationship between the first face and the second face in a graded Peterson's cube of opposition.

Firstly we prove a strong validity of the following syllogisms by concrete syntactical proofs.

Theorem 7. *Syllogisms aEE-I, aBB-I, aDD-I, aGG-I, aVV-I, aZZ-I, and aOO-I are strongly valid in T^{IQ}.*

Proof. Let us assume the syllogism as follows:

$$\textbf{aEE-I:} \quad \frac{(\forall x)(\neg Mx \Rightarrow \neg Px)}{(\forall x)(Sx \Rightarrow \neg Mx)}$$
$$\overline{(\forall x)(Sx \Rightarrow \neg Px).}$$

We know that:

$$T^{IQ} \vdash (\neg Mx \Rightarrow \neg Px) \Rightarrow ((Sx \Rightarrow \neg Mx) \Rightarrow (Sx \Rightarrow \neg Px)).$$

By the rule of generalization of $(\forall x)$ and using the properties of the quantifiers, we have:

$$T^{IQ} \vdash (\forall x)(\neg Mx \Rightarrow \neg Px) \Rightarrow ((\forall x)(Sx \Rightarrow \neg Mx) \Rightarrow (\forall x)(Sx \Rightarrow \neg Px)).$$

□

Proof. Let us assume the syllogism as follows:

$$\textbf{aOO-I:} \quad \frac{(\forall x)(\neg Mx \Rightarrow \neg Px)}{(\exists x)(Sx \wedge \neg Mx)}$$
$$\overline{(\exists x)(Sx \wedge \neg Px).}$$

We know that:

$$T^{IQ} \vdash (\neg Mx \Rightarrow \neg Px) \Rightarrow ((Sx \wedge \neg Mx) \Rightarrow (Sx \wedge \neg Px)).$$

By the rule of generalization $(\forall x)$ and using the properties of the quantifiers, we have:

$$T^{IQ} \vdash (\forall x)(\neg Mx \Rightarrow \neg Px) \Rightarrow ((\exists x)(Sx \wedge \neg Mx) \Rightarrow (\exists x)(Sx \wedge \neg Px)).$$

□

Proof. Let us assume the syllogism as follows:

$$\textbf{aBB-I:} \quad \frac{(\forall x)(\neg Mx \Rightarrow \neg Px)}{(\exists z)[(\forall x)((S|z)x \Rightarrow \neg Mx) \wedge (BiEx)((\mu S)(S|z))]}$$
$$\overline{(\exists z)[(\forall x)((S|z)x \Rightarrow \neg Px) \wedge (BiEx)((\mu S)(S|z))].}$$

Let us denote, by $Ev := (BiEx)((\mu S)(S|z))$.
We know that:

$$T^{IQ} \vdash (\neg Mx \Rightarrow \neg Px) \Rightarrow (((S|z)x \Rightarrow \neg Mx) \Rightarrow ((S|z)x \Rightarrow \neg Px)).$$

By the rule of generalization $(\forall x)$ and using the properties of the quantifiers, we have:

$$T^{IQ} \vdash (\forall x)(\neg Mx \Rightarrow \neg Px) \Rightarrow ((\forall x)((S|z)x \Rightarrow \neg Mx) \Rightarrow ((\forall x)(S|z)x \Rightarrow \neg Px)).$$

By Ł-FTT properties, we have the provable formula as follows:

$$T^{IQ} \vdash (\forall x)(\neg Mx \Rightarrow \neg Px) \Rightarrow$$
$$\Rightarrow \big([(\forall x)((S|z)x \Rightarrow \neg Mx) \wedge Ev] \Rightarrow [(\forall x)((S|z)x \Rightarrow \neg Px) \wedge Ev]\big).$$

Using the generalization rule for $(\forall z)$ and by the properties of the quantifiers we know that:

$$T^{IQ} \vdash (\forall x)(\neg Mx \Rightarrow \neg Px) \Rightarrow$$
$$\Rightarrow \big((\exists z)[(\forall x)((S|z)x \Rightarrow \neg Mx) \wedge Ev] \Rightarrow (\exists z)[(\forall x)((S|z)x \Rightarrow \neg Px) \wedge Ev]\big).$$

By putting $Ev := (BiEx)((\mu S)(S|z))$, we obtain the strong validity of **aBB**-I. If we denote $Ev := (BiVe)((\mu S)(S|z))$, we obtain the strong validity of **aDD**-I. If we put $Ev := (\neg Sm)((\mu S)(S|z))$, we have the strong validity of syllogism **aGG**-I. By denoting $Ev := (SmSi)((\mu S)(S|z))$, we obtain the strong validity of syllogism **aVV**-I. Finally, if we denote $Ev := (SmVe)((\mu S)(S|z))$, we conclude that the syllogism **aZZ**-I is strongly valid. □

From these strongly valid syllogisms, we can obtain other strongly valid syllogisms by using monotonicity. Below, we continue with other forms of valid syllogisms of Figure I.

Theorem 8. *Let syllogisms* **aEE**-*I,* **aBB**-*I,* **aDD**-*I,* **aGG**-*I,* **aVV**-*I,* **aZZ**-*I,* **aOO**-*I be strongly valid in T^{IQ}. Then, the following syllogisms are strongly valid in T^{IQ}:*

aEE						
aEB	aBB					
aED	aBD	aDD				
aEG	aBG	aDG	aGG			
aEV	aBV	aDV	aGV	aVV		
aEZ	aBZ	aDZ	aGZ	aVZ	aZZ	
a(*E)O	a(*B)O	a(*D)O	a(*G)O	a(*V)O	a(*Z)O	aOO

Proof. From strongly valid syllogism **aEE**-I (Theorem 7) and from monotonicity (Theorem A2) by transitivity, we prove the strong validity of syllogisms in the first column. We prove the strong validity of syllogisms in other columns analogously. □

Theorem 9. *Syllogisms* **Aee**-*I,* **Abb**-*I,* **Add**-*I,* **Agg**-*I* **Avv**-*I,* **Azz**-*I,* **Aoo**-*I are strongly valid in T^{IQ}.*

Proof. Analogously to the proof of Theorem 7, this can be proved by replacing each formula by its negation. □

Theorem 10. *Let syllogisms* **Aee**-*I,* **Abb**-*I,* **Add**-*I,* **Agg**-*I,* **Avv**-*I,* **Azz**-*I, and* **Aoo**-*I be strongly valid in T^{IQ}. Then, the following syllogisms are strongly valid in T^{IQ}:*

Aee						
Aeb	Abb					
Aed	Abd	Add				
Aeg	Abg	Adg	Agg			
Aev	Abv	Adv	Agv	Avv		
Aez	Abz	Adz	Agz	Avz	Azz	
A(*e)o	A(*b)o	A(*d)o	A(*g)o	A(*v)o	A(*z)o	Aoo

Proof. This can be proved by monotonicity (Theorem 4) similarly to Theorem 8. □

In other constructions, we will assume a selected group of valid syllogisms (these proofs can be obtained similarly as in Theorem 7), from which we will verify the validity of other forms of syllogisms, especially with the help of monotonicity. In Theorem 11 and in Theorem 12, we present other strongly valid syllogisms of Figure I.

Theorem 11. *Let syllogisms* **Eea**-*I,* **Ebp**-*I,* **Edt**-*I,* **Egk**-*I,* **Evf**-*I,* **Ezs**-*I, and* **Eoi**-*I be strongly valid in* T^{IQ}. *Then, the following syllogisms are strongly valid in* T^{IQ}:

Eea						
Eep	Ebp					
Eet	Ebt	Edt				
Eek	Ebk	Edk	Egk			
Eef	Ebf	Edf	Egf	Evf		
Ees	Ebs	Eds	Egs	Evs	Ezs	
E(*e)i	E(*b)i	E(*d)i	E(*g)i	E(*v)i	E(*z)i	Eoi

Proof. This can be proved by monotonicity (Theorem 4), similarly to Theorem 8. □

Theorem 12. *Let syllogisms* **eEA**-*I,* **eBP**-*I,* **eDT**-*I,* **eGK**-*I,* **eVF**-*I,* **eZS**-*I, and* **eOI**-*I be strongly valid in* T^{IQ}. *Then, the following syllogisms are strongly valid in* T^{IQ}:

eEA						
eEP	eBP					
eET	eBT	eDT				
eEK	eBK	eDK	eGK			
eEF	eBF	eDF	eGF	eVF		
eES	eBS	eDS	eGS	eVS	eZS	
e(*E)I	e(*B)I	e(*D)I	e(*G)I	e(*V)I	e(*Z)I	eOI

Proof. This can be proved by monotonicity (Theorem A2), analogously to Theorem 8. □

We can prove other strongly valid syllogisms using the following proposition, which shows the relationship of the sub-alterns between the first and second squares of opposition.

Proposition 2 ([19]). *The following is provable:*

(a) $T^{IQ} \vdash \mathbf{A} \Rightarrow \mathbf{i}$;
(b) $T^{IQ} \vdash \mathbf{E} \Rightarrow \mathbf{o}$;
(c) $T^{IQ} \vdash \mathbf{a} \Rightarrow \mathbf{I}$;
(d) $T^{IQ} \vdash \mathbf{e} \Rightarrow \mathbf{O}$.

Theorem 13. *Let syllogisms* **AAA**-*I,* **aaa**-*I,* **EAE**-*I,* **eae**-*I,* **Aee**-*I,* **aEE**-*I,* **Eea**-*I, and* **eEA**-*I be strongly valid in* T^{IQ}. *Then, the following syllogisms are strongly valid in* T^{IQ}: **(*A)Ai**-*I* **(*a)aI**-*I,* **(*E)Ao**-*I,* **(*e)aO**-*I,* **(*A)eO**-*I,* **(*a)Eo**-*I,* **(*E)eI**-*I, and* **(*e)Ei**-*I.*

Proof. This can be proved by transitivity and by Proposition 2. □

Theorem 14. *Syllogisms* **oAo**-*I,* **iAi**-*I,* **oeI**-*I,* **ieO**-*I,* **IEo**-*I,* **OEi**-*I,* **IaI**-*I, and* **OaO**-*I are strongly valid in* T^{IQ}.

Proof. Let us assume the syllogism as follows:

$$\mathbf{oAo}\text{-I:} \quad \frac{(\exists x)(\neg Mx \wedge Px)}{(\forall x)(Sx \Rightarrow Mx)}\\ \overline{(\exists x)(\neg Sx \wedge Px)}.$$

We know that:

$$T^{IQ} \vdash (Sx \Rightarrow Mx) \Rightarrow (\neg Mx \Rightarrow \neg Sx).$$

We also know that:
$$T^{IQ} \vdash (\neg Mx \Rightarrow \neg Sx) \Rightarrow ((\neg Mx \wedge Px) \Rightarrow (\neg Sx \wedge Px)).$$

By transitivity, we obtain:
$$T^{IQ} \vdash (Sx \Rightarrow Mx) \Rightarrow ((\neg Mx \wedge Px) \Rightarrow (\neg Sx \wedge Px)).$$

By the rule of generalization for $(\forall x)$ and using the properties of the quantifiers, we have:
$$T^{IQ} \vdash (\forall x)(Sx \Rightarrow Mx) \Rightarrow ((\exists x)(\neg Mx \wedge Px) \Rightarrow (\exists x)(\neg Sx \wedge Px)).$$

By adjunction, we obtain:
$$T^{IQ} \vdash ((\exists x)(\neg Mx \wedge Px) \& (\forall x)(Sx \Rightarrow Mx)) \Rightarrow (\exists x)(\neg Sx \wedge Px).$$

The strong validity of syllogism **OaO**-I can be proven analogously, by replacing each formula with its negation. The strong validity of other syllogisms can be proven similarly. □

Theorem 15. *Let syllogisms* **oAo**-*I,* **iAi**-*I,* **oeI**-*I,* **ieO**-*I,* **IEo**-*I,* **OEi**-*I,* **IaI**-*I, and* **OaO**-*I be strongly valid in* T^{IQ}. *Then, the following syllogisms are strongly valid in* T^{IQ}:

(*e)Ao	(*E)aO	(*a)Ai	(*A)aI	(*e)eI	(*E)Ei	(*a)eO	(*A)Eo
(*b)Ao	(*B)aO	(*p)Ai	(*P)aI	(*b)eI	(*B)Ei	(*p)eO	(*P)Eo
(*d)Ao	(*D)aO	(*t)Ai	(*T)aI	(*d)eI	(*D)Ei	(*t)eO	(*T)Eo
(*g)Ao	(*G)aO	(*k)Ai	(*K)aI	(*g)eI	(*G)Ei	(*k)eO	(*K)Eo
(*v)Ao	(*V)aO	(*f)Ai	(*F)aI	(*v)eI	(*V)Ei	(*f)eO	(*F)Eo
(*z)Ao	(*Z)aO	(*s)Ai	(*S)aI	(*z)eI	(*Z)Ei	(*s)eO	(*S)Eo
oAo	OaO	iAi	IaI	oeI	OEi	ieO	IEo

Proof. By monotonicity (Theorem 4) and the strongly valid syllogism **oAo**-I, we prove, by transitivity, the strong validity of the syllogisms in the first column. We prove the other syllogisms in the other columns analogously by monotonicity (Theorems A2 and 4). □

3.5. New Forms of Figure II

Figure II is similar to Figure I, so we will not present concrete syntactical proofs of syllogisms. The syllogisms can be proved similarly to Theorems 7 and 14.

Theorem 16. *Let syllogisms* **aAA**-*II,* **aPP**-*II,* **aTT**-*II,* **aKK**-*II,* **aFF**-*II,* **aSS**-*II, and* **aII**-*II be strongly valid in* T^{IQ}. *Then, the following syllogisms are strongly valid in* T^{IQ}:

aAA						
aAP	aPP					
aAT	aPT	aTT				
aAK	aPK	aTK	aKK			
aAF	aPF	aTF	aKF	aFF		
aAS	aPS	aTS	aKS	aFS	aSS	
a(*A)I	a(*P)I	a(*T)I	a(*K)I	a(*F)I	a(*S)I	aII

Proof. This can be proved by monotonicity (Theorem A2), similarly to Theorem 8. □

Theorem 17. *Let syllogisms* **Aaa**-*II,* **App**-*II,* **Att**-*II,* **Akk**-*II,* **Aff**-*II,* **Ass**-*II, and* **Aii**-*II be strongly valid in* T^{IQ}. *Then, the following syllogisms are strongly valid in* T^{IQ}:

$$\begin{array}{lllllll}
Aaa \\
Aap & App \\
Aat & Apt & Att \\
Aak & Apk & Atk & Akk \\
Aaf & Apf & Atf & Akf & Aff \\
Aas & Aps & Ats & Aks & Afs & Ass \\
A(*a)i & A(*p)i & A(*t)i & A(*k)i & A(*f)i & A(*s)i & Aii
\end{array}$$

Proof. This can be proved by monotonicity (Theorem 4), similarly to Theorem 8. □

Theorem 18. *Let syllogisms* **eEA**-*II,* **eBP**-*II,* **eDT**-*II,* **eGK**-*II,* **eVF**-*II,* **eZS**-*II, and* **eOI**-*II be strongly valid in* T^{IQ}. *Then, the following syllogisms are strongly valid in* T^{IQ}:

$$\begin{array}{lllllll}
eEA \\
eEP & eBP \\
eET & eBT & eDT \\
eEK & eBK & eDK & eGK \\
eEF & eBF & eDF & eGF & eVF \\
eES & eBS & eDS & eGS & eVS & eZS \\
e(*E)I & e(*B)I & e(*D)I & e(*G)I & e(*V)I & e(*Z)I & eOI
\end{array}$$

Proof. This can be proved by monotonicity (Theorem A2), similarly to Theorem 8. □

Theorem 19. *Let syllogisms* **Eea**-*II,* **Ebp**-*II,* **Edt**-*II,* **Egk**-*II,* **Evf**-*II,* **Ezs**-*II, and* **Eoi**-*II be strongly valid in* T^{IQ}. *Then, the following syllogisms are strongly valid in* T^{IQ}:

$$\begin{array}{lllllll}
Eea \\
Eep & Ebp \\
Eet & Ebt & Edt \\
Eek & Ebk & Edk & Egk \\
Eef & Ebf & Edf & Egf & Evf \\
Ees & Ebs & Eds & Egs & Evs & Ezs \\
E(*e)i & E(*b)i & E(*d)i & E(*g)i & E(*v)i & E(*z)i & Eoi
\end{array}$$

Proof. This can be proved by monotonicity (Theorem 4), similarly to Theorem 8. □

Theorem 20. *Let syllogisms* **EAE**-*II,* **eae**-*II,* **AEE**-*II,* **aee**-*II,* **aAA**-*II,* **Aaa**-*II,* **Eea**-*II, and* **eEA**-*II be strongly valid in* T^{IQ}. *Then, the following syllogisms are strongly valid in* T^{IQ}: **(*E)Ao**-*II,* **(*e)aO**-*II,* **(*A)Eo**-*II,* **(*a)eO**-*II,* **(*a)Ai**-*II,* **(*A)aI**-*II,* **(*E)eI**-*II, and* **(*e)Ei**-*II.*

Proof. This can be proved by transitivity and by Proposition 2. □

Theorem 21. *Let syllogisms* **OAo**-*II,* **oaO**-*II,* **IEo**-*II,* **ieO**-*II,* **OeI**-*II,* **oEi**-*II,* **IaI**-*II, and* **iAi**-*II be strongly valid in* T^{IQ}. *Then, the following syllogisms are strongly valid in* T^{IQ}:

$$\begin{array}{lllllll}
(*E)Ao & (*e)aO & (*A)Eo & (*a)eO & (*E)eI & (*e)Ei & (*a)Ai & (*A)aI \\
(*B)Ao & (*b)aO & (*P)Eo & (*p)eO & (*B)eI & (*b)Ei & (*p)Ai & (*P)aI \\
(*D)Ao & (*d)aO & (*T)Eo & (*t)eO & (*D)eI & (*d)Ei & (*t)Ai & (*T)aI \\
(*G)Ao & (*g)aO & (*K)Eo & (*k)eO & (*G)eI & (*g)Ei & (*k)Ai & (*K)aI \\
(*V)Ao & (*v)aO & (*F)Eo & (*f)eO & (*V)eI & (*v)Ei & (*f)Ai & (*F)aI \\
(*Z)Ao & (*z)aO & (*S)Eo & (*s)eO & (*Z)eI & (*z)Ei & (*s)Ai & (*S)aI \\
OAo & oaO & IEo & ieO & OeI & oEi & iAi & IaI
\end{array}$$

Proof. This can be proved by monotonicity (Theorems A2 and 4), analogously to Theorem 15. □

Theorem 22. *Syllogisms **(*A)Ai**-II, **(*a)aI**-II, **(*E)Ei**-II, **(*e)eI**-II, **(*e)Ao**-II, **(*E)aO**-II, **(*A)eO**-II, and **(*a)Eo**-II are strongly valid in T^{IQ}.*

Proof. Let us assume the syllogism as follows:

$$\textbf{(*A)Ai-II:}\quad \frac{(\forall x)(Px \Rightarrow Mx)\,\&\,(\exists x)(\neg Mx)}{(\forall x)(Sx \Rightarrow Mx)}$$
$$\overline{(\exists x)(\neg Sx \wedge \neg Px).}$$

We know that:
$$T^{IQ} \vdash (Sx \Rightarrow Mx) \Rightarrow (\neg Mx \Rightarrow \neg Sx). \tag{14}$$

We know that:
$$T^{IQ} \vdash (Px \Rightarrow Mx) \Rightarrow (\neg Mx \Rightarrow \neg Px). \tag{15}$$

We also know that:
$$T^{IQ} \vdash (\neg Mx \Rightarrow \neg Sx) \Rightarrow ((\neg Mx \Rightarrow \neg Px) \Rightarrow (\neg Mx \Rightarrow (\neg Sx \wedge \neg Px))). \tag{16}$$

By the application of transitivity on (14) and (16), we obtain:
$$T^{IQ} \vdash (Sx \Rightarrow Mx) \Rightarrow ((\neg Mx \Rightarrow \neg Px) \Rightarrow (\neg Mx \Rightarrow (\neg Sx \wedge \neg Px))).$$

By adjunction, we obtain:
$$T^{IQ} \vdash (\neg Mx \Rightarrow \neg Px) \Rightarrow ((Sx \Rightarrow Mx) \Rightarrow (\neg Mx \Rightarrow (\neg Sx \wedge \neg Px))). \tag{17}$$

By the application of transitivity on (15) and (17), we obtain:
$$T^{IQ} \vdash (Px \Rightarrow Mx) \Rightarrow ((Sx \Rightarrow Mx) \Rightarrow (\neg Mx \Rightarrow (\neg Sx \wedge \neg Px))).$$

By generalization $(\forall x)$ and the properties of the quantifiers, we obtain:
$$T^{IQ} \vdash (\forall x)(Px \Rightarrow Mx) \Rightarrow ((\forall x)(Sx \Rightarrow Mx) \Rightarrow ((\exists x)\neg Mx \Rightarrow (\exists x)(\neg Sx \wedge \neg Px))).$$

The strong validity of other fuzzy logical syllogisms can be verified similarly. □

3.6. New Forms of Figure III

Firstly, we will show some syntactical proofs of syllogisms on this Figure.

Theorem 23. *Syllogisms **AOo**-III, ***(PG)o**-III, ***(TD)o**-III, ***(KB)o**-III, and **IEo**-III are strongly valid in T^{IQ}.*

Proof. Let us assume the syllogism as follows:

$$\textbf{AOo-III:}\quad \frac{(\forall x)(Mx \Rightarrow Px)}{(\exists x)(Mx \wedge \neg Sx)}$$
$$\overline{(\exists x)(\neg Sx \wedge Px).}$$

We know that:
$$T^{IQ} \vdash (Mx \Rightarrow Px) \Rightarrow ((Mx \wedge \neg Sx) \Rightarrow (\neg Sx \wedge Px)).$$

Analogously, as in previous proofs, using the properties of the quantifiers, we obtain:
$$T^{IQ} \vdash (\forall x)(Mx \Rightarrow Px) \Rightarrow ((\exists x)(Mx \wedge \neg Sx) \Rightarrow (\exists x)(\neg Sx \wedge Px)).$$

□

Proof. Let us assume the syllogism as follows:

IEo-III: $\dfrac{(\exists x)(Mx \wedge Px)}{(\forall x)(Mx \Rightarrow \neg Sx)}$
$\overline{(\exists x)(\neg Sx \wedge Px).}$

We know that:

$$T^{IQ} \vdash (Mx \Rightarrow \neg Sx) \Rightarrow ((Mx \wedge Px) \Rightarrow (\neg Sx \wedge Px)).$$

Using the generalization of $(\forall x)$ and by the logical properties of the quantifiers, we have:

$$T^{IQ} \vdash (\forall x)(Mx \Rightarrow \neg Sx) \Rightarrow ((\exists x)(Mx \wedge Px) \Rightarrow (\exists x)(\neg Sx \wedge Px)).$$

By adjunction, we obtain:

$$T^{IQ} \vdash (\exists x)(Mx \wedge Px) \Rightarrow ((\forall x)(Mx \Rightarrow \neg Sx) \Rightarrow (\exists x)(\neg Sx \wedge Px)).$$

□

Proof. Let us assume the syllogism as follows:

PGo-III: $\dfrac{(\exists z)[(\forall x)((M|z)x \Rightarrow Px) \wedge (BiEx)((\mu M)(M|z))]}{(\exists z')[(\forall x)((M|z')x \Rightarrow \neg Sx) \wedge (\neg Sm)((\mu M)(M|z'))]}$
$\overline{(\exists x)(\neg Sx \wedge Px).}$

We know that:

$$T^{IQ} \vdash ((M|z)x \Rightarrow Px) \Rightarrow$$
$$\Rightarrow (((M|z')x \Rightarrow \neg Sx) \Rightarrow (((M|z)x \& (M|z')x) \Rightarrow (\neg Sx \wedge Px))).$$

By generalization $(\forall x)$ and the properties of the quantifiers, we obtain:

$$T^{IQ} \vdash (\forall x)((M|z)x \Rightarrow Px) \Rightarrow$$
$$\Rightarrow ((\forall x)((M|z')x \Rightarrow \neg Sx) \Rightarrow ((\exists x)((M|z)x \& (M|z')x) \Rightarrow (\exists x)(\neg Sx \wedge Px))).$$

By using the property of Łukasiewicz logic, we obtain:

$$T^{IQ} \vdash [(\forall x)((M|z)x \Rightarrow Px) \wedge Ev] \Rightarrow$$
$$\Rightarrow ((\forall x)((M|z')x \Rightarrow \neg Sx) \Rightarrow ((\exists x)((M|z)x \& (M|z')x) \Rightarrow (\exists x)(\neg Sx \wedge Px))).$$

By adjunction and the property of Łukasiewicz logic, we obtain:

$$T^{IQ} \vdash [(\forall x)((M|z')x \Rightarrow \neg Sx) \wedge Ev'] \Rightarrow$$
$$\Rightarrow ([(\forall x)((M|z)x \Rightarrow Px) \wedge Ev] \Rightarrow ((\exists x)((M|z)x \& (M|z')x) \Rightarrow (\exists x)(\neg Sx \wedge Px))).$$

By adjunction, we obtain:

$$T^{IQ} \vdash ([(\forall x)((M|z)x \Rightarrow Px) \wedge Ev] \& [(\forall x)((M|z')x \Rightarrow \neg Sx) \wedge Ev'] \&$$
$$\& (\exists x)((M|z)x \& (M|z')x)) \Rightarrow (\exists x)(\neg Sx \wedge Px).$$

By generalization $(\forall z)$, $(\forall z')$, and the properties of the quantifiers, we obtain:

$$T^{IQ} \vdash (\exists z)(\exists z')([(\forall x)((M|z)x \Rightarrow Px) \wedge Ev] \& [(\forall x)((M|z')x \Rightarrow \neg Sx) \wedge Ev'] \&$$
$$\& (\exists x)((M|z)x \& (M|z')x)) \Rightarrow (\exists x)(\neg Sx \wedge Px).$$

If we put $Ev := (BiEx)((\mu M)(M|z))$ and $Ev' := (\neg Sm)((\mu M)(M|z'))$, we obtain the strong validity of syllogism *(**PG**)o-III. If we put $Ev := (BiVe)((\mu M)(M|z))$ and

$Ev' := (BiVe)((\mu M)(M|z'))$, we obtain the strong validity of syllogism ***(TD)o**-III. If we put $Ev := (\neg Sm)((\mu M)(M|z))$ and $Ev' := (BiEx)((\mu M)(M|z'))$, we obtain the strong validity of syllogism ***(KB)o**-III. □

Theorem 24. *Let syllogisms* **AOo**-*III,* ***(PG)o**-*III,* ***(TD)o**-*III,* ***(KB)o**-*III, and* **IEo**-*III be strongly valid in* T^{IQ}. *Then, the following syllogisms are strongly valid in* T^{IQ}:

(*A)Eo	(*P)Eo	(*T)Eo	(*K)Eo	(*F)Eo	(*S)Eo	IEo
A(*B)o	*(PB)o	*(TB)o	*(KB)o			
A(*D)o	*(PD)o	*(TD)o				
A(*G)o	*(PG)o					
A(*V)o						
A(*Z)o						
AOo						

Proof. From the strongly valid logical syllogism **AOo**-III, and using the monotonicity (Theorem A2), we prove the strong validity of the syllogisms in the first column by transitivity. From the strongly valid syllogism **IEo**-III, and by monotonicity (Theorem A2), we can verify the strong validity of the syllogisms in the first row by transitivity. Analogously, using the strongly valid syllogism ***(PG)o**-III and by monotonicity (Theorem A2), we can verify the strong validity of the syllogisms in the second column by transitivity. The syllogisms in the third and fourth column can be proven analogously. □

Theorem 25. *Syllogisms* **aoO**-*III,* ***(pg)O**-*III,* ***(td)O**-*III,* ***(kb)O**-*III, and* **ieO**-*III are strongly valid in* T^{IQ}.

Proof. This can be proven analogously to the proof of Theorem 23, by replacing each formula with its negation. □

Theorem 26. *Let syllogisms* **aoO**-*III,* ***(pg)O**-*III,* ***(td)O**-*III,* ***(kb)O**-*III, and* **ieO**-*III be strongly valid in* T^{IQ}. *Then, the following syllogisms are strongly valid in* T^{IQ}:

(*a)eO	(*p)eO	(*t)eO	(*k)eO	(*f)eO	(*s)eO	ieO
a(*b)O	*(pb)O	*(tb)O	*(kb)O			
a(*d)O	*(pd)O	*(td)O				
a(*g)O	*(pg)O					
a(*v)O						
a(*z)O						
aoO						

Proof. This can be proven by monotonicity (Theorem 4), similarly to Theorem 24. □

Next, we will consider the strong validity of some syllogisms without concrete proofs. These proofs are similar to the proofs in Theorem 23.

Theorem 27. *Let syllogisms* **EOi**-*III,* ***(BG)i**-*III,* ***(DD)i**-*III,* ***(GB)i**-*III, and* **OEi**-*III be strongly valid in* T^{IQ}. *Then, the following syllogisms are strongly valid in* T^{IQ}:

E(*E)i	(*B)Ei	(*D)Ei	(*G)Ei	(*V)Ei	(*Z)Ei	OEi
E(*B)i	*(BB)i	*(DB)i	*(GB)i			
E(*D)i	*(BD)i	*(DD)i				
E(*G)i	*(BG)i					
E(*V)i						
E(*Z)i						
EOi						

Proof. This can be proven by monotonicity (Theorem A2), similarly to Theorem 24. □

Theorem 28. *Let syllogisms* **oeI**-*III,* ***(gb)I**-*III,* ***(dd)I**-*III,* ***(bg)I**-*III, and* **eoI**-*III be strongly valid in T^{IQ}. Then, the following syllogisms are strongly valid in T^{IQ}:*

e(*e)I	(*b)eI	(*d)eI	(*g)eI	(*v)eI	(*z)eI	oeI
e(*b)I	*(bb)I	*(db)I	*(gb)I		
e(*d)I	*(bd)I	*(dd)I			
e(*g)I	*(bg)I				
e(*v)I					
e(*z)I					
eoI					

Proof. This can be proven by monotonicity (Theorem 4), similarly to Theorem 24. □

The following syllogisms are also strongly valid in Figure III.

Theorem 29. *Let syllogisms* **eAe**-*III,* **EaE**-*III,* **aAa**-*III,* **AaA**-*III,* **eEA**-*III,* **Eea**-*III,* **aEE**-*III, and* **Aee**-*III be strongly valid in T^{IQ}. Then, the following syllogisms are strongly valid in T^{IQ}:*

eAe	EaE	aAa	AaA	eEA	Eea	aEE	Aee
eAb	EaB	aAp	AaP	eEP	Eep	aEB	Aeb
eAd	EaD	aAt	AaT	eET	Eet	aED	Aed
eAg	EaG	aAk	AaK	eEK	Eek	aEG	Aeg
eAv	EaV	aAf	AaF	eEF	Eef	aEV	Aev
eAz	EaZ	aAs	AaS	eES	Ees	aEZ	Aez
e(*A)o	E(*a)O	a(*A)i	A(*a)I	e(*E)I	E(*e)i	a(*E)O	A(*e)o

Proof. From the strongly valid syllogism **eAe**-III, and from monotonicity (Theorem 4), we can prove, by transitivity, the strong validity of the syllogisms in the first column. Analogously, from monotonicity (Theorems A2 and 4), we can prove the strong validity of the syllogisms in the other columns. □

Theorem 30. *Let syllogisms* **eAe**-*III,* **EaE**-*III,* **aAa**-*III,* **AaA**-*III,* **eEA**-*III,* **Eea**-*III,* **aEE**-*III, and* **Aee**-*III be strongly valid in T^{IQ}. Then, the following syllogisms are strongly valid in T^{IQ}:* ***(*e)AO**-*III,* ***(*E)ao**-*III,* ***(*a)AI**-*III,* ***(*A)ai**-*III,* ***(*e)Ei**-*III,* ***(*E)eI**-*III,* ***(*a)Eo**-*III, and* ***(*A)eO**-*III.*

Proof. This can be proven by transitivity and by Proposition 2. □

3.7. New Forms of Figure IV

Firstly, we show a proof of the strongly valid syllogisms of this figure.

Theorem 31. *Syllogisms* **aII**-*IV,* **Aii**-*IV,* **EOi**-*IV,* **eoI**-*IV,* **aOo**-*IV, and* **AoO**-*IV are strongly valid in T^{IQ}.*

Proof. Let us assume the syllogism as follows:

$$\textbf{aII-IV:} \quad \frac{(\forall x)(\neg Px \Rightarrow \neg Mx)}{(\exists x)(Mx \wedge Sx)}$$
$$\overline{(\exists x)(Sx \wedge Px).}$$

We know that:
$$T^{IQ} \vdash (\neg Px \Rightarrow \neg Mx) \Rightarrow (Mx \Rightarrow Px). \tag{18}$$

We also know that:
$$T^{IQ} \vdash (Mx \Rightarrow Px) \Rightarrow ((Mx \wedge Sx) \Rightarrow (Sx \wedge Px)). \tag{19}$$

By the application of transitivity on (18) and (19), we obtain:

$$T^{IQ} \vdash (\neg Px \Rightarrow \neg Mx) \Rightarrow ((Mx \wedge Sx) \Rightarrow (Sx \wedge Px)).$$

By generalization ($\forall x$) and quatifier properties, we obtain:

$$T^{IQ} \vdash (\forall x)(\neg Px \Rightarrow \neg Mx) \Rightarrow ((\exists x)(Mx \wedge Sx) \Rightarrow (\exists x)(Sx \wedge Px)).$$

The strong validity of other syllogisms can be proven similarly. □

Theorem 32. *Let syllogisms* **aII**-*IV,* **Aii**-*IV,* **EOi**-*IV,* **eoI**-*IV,* **aOo**-*IV, and* **AoO**-*IV be strongly valid in* T^{IQ}. *Then, the following syllogisms are strongly valid in* T^{IQ}:

a(*A)I	A(*a)i	E(*E)i	e(*e)I	a(*E)o	A(*e)O
a(*P)I	A(*p)i	E(*B)i	e(*b)I	a(*B)o	A(*b)O
a(*T)I	A(*t)i	E(*D)i	e(*d)I	a(*D)o	A(*d)O
a(*K)I	A(*k)i	E(*G)i	e(*g)I	a(*G)o	A(*g)O
a(*F)I	A(*f)i	E(*V)i	e(*v)I	a(*V)o	A(*v)O
a(*S)I	A(*s)i	E(*Z)i	e(*z)I	a(*Z)o	A(*z)O
aII	Aii	EOi	eoI	aOo	AoO

Proof. From the strongly valid syllogism **aII**-IV, and from monotonicity (Theorem A2), we prove the strongly valid syllogisms in the first column by transitivity. We can prove the syllogisms in the other columns analogously by using monotonicity (Theorem A2, Theorem 4). □

We will consider the strong validity of some syllogisms without concrete proofs. These proofs are similar to the previous proofs in this article.

Theorem 33. *Let syllogisms* **AAa**-*IV,* **aaA**-*IV,* **eAe**-*IV,* **EaE**-*IV,* **eEA**-*IV, and* **Eea**-*IV be strongly valid in* T^{IQ}. *Then, the following syllogisms are strongly valid in* T^{IQ}:

AAa	aaA	eAe	EaE	eEA	Eea
AAp	aaP	eAb	EaB	eEP	Eep
AAt	aaT	eAd	EaD	eET	Eet
AAk	aaK	eAg	EaG	eEK	Eek
AAf	aaF	eAv	EaV	eEF	Eef
AAs	aaS	eAz	EaZ	eES	Ees
A(*A)i	a(*a)I	e(*A)o	E(*a)O	e(*E)I	E(*e)i

Proof. From the strongly valid syllogism **AAa**-IV and monotonicity (Theorem 4), we prove, by transitivity, the strongly valid syllogism in the first column. Similarly, we can prove the strong validity of the syllogisms in the other columns by monotonicity (Theorems A2 and 4). □

Theorem 34. *Let syllogisms* **eAe**-*IV,* **EaE**-*IV,* **eEA**-*IV, and* **Eea**-*IV be strongly valid in* T^{IQ}. *Then, the following syllogisms are strongly valid in* T^{IQ}: **(*e)AO**-*IV,* **(*E)ao**-*IV,* **(*e)Ei**-*IV, and* **(*E)eI**-*IV.*

Proof. This can be proven by transitivity and by Proposition 2. □

Theorem 35. *Let syllogisms* **oAO**-*IV,* **Oao**-*IV,* **oEi**-*IV,* **OeI**-*IV,* **IEo**-*IV, and* **ieO**-*IV be strongly valid in* T^{IQ}. *Then, the following syllogisms are strongly valid in* T^{IQ}:

(*e)AO	(*E)ao	(*e)Ei	(*E)eI	(*A)Eo	(*a)eO
(*b)AO	(*B)ao	(*b)Ei	(*B)eI	(*P)Eo	(*p)eO
(*d)AO	(*D)ao	(*d)Ei	(*D)eI	(*T)Eo	(*t)eO
(*g)AO	(*G)ao	(*g)Ei	(*G)eI	(*K)Eo	(*k)eO
(*v)AO	(*V)ao	(*v)Ei	(*V)eI	(*F)Eo	(*f)eO
(*z)AO	(*Z)ao	(*z)Ei	(*Z)eI	(*S)Eo	(*s)eO
oAO	Oao	oEi	OeI	IEo	ieO

Proof. From the strongly valid syllogism **oAO**-IV and monotonicity (Theorem 4), we prove the strongly valid syllogisms in the first column by transitivity. The syllogisms in the other columns can be proven similarly, by using monotonicity (Theorems A2 and 4). □

3.8. Examples of Logical Syllogisms in Finite Model

The model theory deals with the relation between syntax and semantics. In finite model theory, an interpretation is restricted to finite structures which have finite universes. In our examples, the models are restricted to finite universes—the universe in the first example consists of six elements and the universe in the second example consists of eight elements.

Below, we introduce several examples of new forms of logical syllogisms. We will show examples of valid syllogisms in the simple model with the finite set M_ϵ of elements. Details of the constructed model can be found in [16]. The built model is $\mathcal{M} = \langle (M_\alpha, =_\alpha)_{\alpha \in Types}, \mathcal{L}_\Delta \rangle$, where $M_o = [0,1]$ is based on the standard Łukasiewicz MV_Δ-algebra. The Łukasiewicz biresiduation $\leftrightarrow$ represents the fuzzy equality $=_o$. The logical implication is represented by the Łukasiewicz residuation $\rightarrow$.

We will assume the example of the fuzzy measure which was proposed in Section 2.3 by Equation (6). For model $\mathcal{M}$, it holds true that $\mathcal{M} \models T^{IQ}$. The syllogism in model $\mathcal{M}$ is valid if

$$\mathcal{M}(P_1) \otimes \mathcal{M}(P_2) \leq \mathcal{M}(C).$$

3.8.1. Example of Valid Syllogism of Figure III

Let us assume the following syllogism:

AOo: P_1 : All flu are viral diseases.
P_2 : Some flu are not diseases transmittable to humans
C Some diseases which are not transmittable to humans are viral diseases.

We assume the frame which is described in the previous subsection. Let M_ϵ be a set of diseases. We consider six diseases $U = \{u_1, u_2, u_3, u_4, u_5, u_6\}$. We interpret the formulas in the considered model as follows: Let $\text{Flu}_{o\epsilon}$ be the formula "flu", with the interpretation $\mathcal{M}(\text{Flu}_{o\epsilon}) = F \subseteq M_\epsilon$ defined by

$$F = \{0.3/u_1, 0.8/u_2, 0.4/u_3, 0.7/u_4, 0.4/u_5, 0.5/u_6\}.$$

Let $\text{Vir}_{o\epsilon}$ be the formula "viral diseases", with the interpretation $\mathcal{M}(\text{Vir}_{o\epsilon}) = V \subseteq M_\epsilon$ defined by

$$V = \{0.8/u_1, 0.2/u_2, 0.3/u_3, 0.9/u_4, 0.7/u_5, 0.6/u_6\}.$$

Let $\text{Tran}_{o\epsilon}$ be the formula "diseases transmittable to humans", with the interpretation $\mathcal{M}(\text{Tran}_{o\epsilon}) = T \subseteq M_\epsilon$ defined by

$$T = \{0.2/u_1, 0.1/u_2, 0.7/u_3, 0.3/u_4, 0.1/u_5, 0.6/u_6\}.$$

Let $\text{Ntran}_{o\epsilon}$ be the formula "diseases not transmittable to humans", with the interpretation $\mathcal{M}(\text{Ntran}_{o\epsilon}) = \neg T \subseteq M_\epsilon$ defined by

$$\neg T = \{0.8/u_1, 0.9/u_2, 0.3/u_3, 0.7/u_4, 0.9/u_5, 0.4/u_6\}.$$

Major premise: "All flu are viral diseases" is the formula:

$$Q_{Bi\Delta}^{\forall}(\text{Flu}_{o\epsilon}, \text{Vir}_{o\epsilon}) := (\forall x_\epsilon)(\text{Flu}_{o\epsilon}(x_\epsilon) \Rightarrow \text{Vir}_{o\epsilon}(x_\epsilon)),$$

which is interpreted by:

$$\mathcal{M}(Q_{Bi\Delta}^{\forall}(\text{Flu}_{o\epsilon}, \text{Vir}_{o\epsilon})) = \bigwedge_{m \in M_\epsilon} (\mathcal{M}(\text{Flu}_{o\epsilon}(m)) \to \mathcal{M}(\text{Vir}_{o\epsilon}(m))) = 0.4. \quad (20)$$

Minor premise: "Some flu are not diseases transmittable to humans" is the formula:

$$Q_{Bi\Delta}^{\exists}(\text{Flu}_{o\epsilon}, \text{Ntran}_{o\epsilon}) := (\exists x_\epsilon)(\text{Flu}_{o\epsilon}(x_\epsilon) \wedge \text{Ntran}_{o\epsilon}(x_\epsilon)),$$

which is interpreted by:

$$\mathcal{M}(Q_{Bi\Delta}^{\exists}(\text{Flu}_{o\epsilon}, \text{Ntran}_{o\epsilon})) = \bigvee_{m \in M_\epsilon} (\mathcal{M}(\text{Flu}_{o\epsilon}(m)) \wedge \mathcal{M}(\text{Ntran}_{o\epsilon}(m))) = 0.8. \quad (21)$$

Conclusion: "Some diseases which are not transmittable to humans are viral diseases" is the formula:

$$Q_{Bi\Delta}^{\exists}(\text{Ntran}_{o\epsilon}, \text{Vir}_{o\epsilon}) := (\exists x_\epsilon)(\text{Ntran}_{o\epsilon}(x_\epsilon) \wedge \text{Vir}_{o\epsilon}(x_\epsilon)),$$

which is interpreted by:

$$\mathcal{M}(Q_{Bi\Delta}^{\exists}(\text{Ntran}_{o\epsilon}, \text{Vir}_{o\epsilon})) = \bigvee_{m \in M_\epsilon} (\mathcal{M}(\text{Ntran}_{o\epsilon}(m)) \wedge \mathcal{M}(\text{Vir}_{o\epsilon}(m))) = 0.8. \quad (22)$$

From (20)–(22), we can see that the condition of validity in our model is satisfied because $\mathcal{M}(P_1) \otimes \mathcal{M}(P_2) = 0.4 \otimes 0.8 = 0.2 \leq \mathcal{M}(C) = 0.8$.

3.8.2. Example of Valid Syllogism of Figure IV

(*g)Ei:
P_1: Many diseases which are not lethal are virus diseases.
P_2: All virus diseases can not be cured by antibiotics.
P_3: Some diseases which can not be cured by antibiotics are not lethal diseases.

We suppose the same frame and the fuzzy measure as in the previous example. Let M_ϵ be a set of diseases. We consider eight diseases $V = \{v_1, v_2, v_3, v_4, v_5, v_6, v_7, v_8\}$. We interpret the formulas in the considered model as follows: Let $\text{Nlethal}_{o\epsilon}$ be the formula "diseases which are not lethal", with the interpretation $\mathcal{M}(\text{Nlethal}_{o\epsilon}) = \neg L \subseteq_\sim M_\epsilon$ defined by

$$\neg L = \{1/v_1, 0.8/v_2, 0.3/v_3, 0.5/v_4, 0.7/v_5, 0.4/v_6, 0.4/v_7, 0.4/v_8\}.$$

Let $\text{Vir}_{o\epsilon}$ be the formula "virus diseases", with the interpretation $\mathcal{M}(\text{Vir}_{o\epsilon}) = V \subseteq_\sim M_\epsilon$ defined by

$$V = \{0.6/v_1, 0.4/v_2, 0.2/v_3, 0.6/v_4, 0.7/v_5, 0.5/v_6, 0.3/v_7, 0.2/v_8\}.$$

Let $\text{Natb}_{o\epsilon}$ be the formula "diseases which cannot be cured by antibiotics", with the interpretation $\mathcal{M}(\text{Natb}_{o\epsilon}) = \neg A \subseteq_\sim M_\epsilon$ defined by

$$\neg A = \{0.7/v_1, 0.5/v_2, 0.3/v_3, 0.7/v_4, 0.7/v_5, 0.5/v_6, 0.3/v_7, 0.5/v_8\}.$$

Major premise: "Many diseases which are not lethal are virus diseases." can be represented in our model as:

$$\mathcal{M}((\exists z_{o\epsilon})[(\forall x_\epsilon)((\text{Nlethal}_{o\epsilon}|z_{o\epsilon})(x_\epsilon) \Rightarrow \text{Vir}_{o\epsilon}(x_\epsilon)) \&$$
$$\& (\exists x_\epsilon)(\text{Nlethal}_{o\epsilon}|z_{o\epsilon})(x_\epsilon) \wedge (\neg Sm)((\mu\,\text{Nlethal}_{o\epsilon})(\text{Nlethal}_{o\epsilon}|z_{o\epsilon}))]). \quad (23)$$

This leads us to find the fuzzy set $\mathcal{M}(\text{Nlethal}_{o\epsilon}|z_{o\epsilon}) = C \subseteq M_\epsilon$, which gives us the greatest degree in (23). It can be confirmed that fuzzy set $C = \{0.5/v_4, 0.7/v_5, 0.4/v_6\} \subseteq \neg L$ leads to the greatest degree in (23).

$$\mathcal{M}(Q_{\neg Sm}^\forall(\text{Nlethal}_{o\epsilon}, \text{Vir}_{o\epsilon})) = 1 \otimes 0.7 \wedge 1 = 0.7. \tag{24}$$

Minor premise: "All virus diseases can not be cured by antibiotics" is the formula:

$$Q_{Bi\Delta}^\forall(\text{Vir}_{o\epsilon}, \text{Natb}_{o\epsilon}) := (\forall x_\epsilon)(\text{Vir}_{o\epsilon}(x_\epsilon) \Rightarrow \text{Natb}_{o\epsilon}(x_\epsilon)),$$

which is interpreted by:

$$\mathcal{M}(Q_{Bi\Delta}^\forall(\text{Vir}_{o\epsilon}, \text{Natb}_{o\epsilon})) = \bigwedge_{m \in M_\epsilon} (\mathcal{M}(\text{Vir}_{o\epsilon}(m)) \to \mathcal{M}(\text{Natb}_{o\epsilon}(m))) = 1. \tag{25}$$

Conclusion: "Some diseases which can not be cured by antibiotics are not lethal diseases" is the formula:

$$Q_{Bi\Delta}^\exists(\text{Natb}_{o\epsilon}, \text{Nlethal}_{o\epsilon}) := (\exists x_\epsilon)(\text{Natb}_{o\epsilon}(x_\epsilon) \wedge \text{Nlethal}_{o\epsilon}(x_\epsilon)),$$

which is interpreted by:

$$\mathcal{M}(Q_{Bi\Delta}^\exists(\text{Natb}_{o\epsilon}, \text{Nlethal}_{o\epsilon})) = \bigvee_{m \in M_\epsilon} (\mathcal{M}(\text{Natb}_{o\epsilon}(m)) \wedge \mathcal{M}(\text{Nlethal}_{o\epsilon}(m))) = 0.7. \tag{26}$$

From (24)–(26), we can see that the condition of validity is satisfied in our model because $\mathcal{M}(P_1) \otimes \mathcal{M}(P_2) = 0.7 \otimes 1 = 0.7 \leq \mathcal{M}(C) = 0.7$.

4. Discussion

In the discussion section, we will comment on new forms of fuzzy syllogisms which we have proven in Section 3. In Section 3, we can see that we can order strongly valid syllogisms into triangles or columns by monotonicity. In the vertexes of these triangles are fuzzy syllogisms that consist of classical quantifiers or new forms of classical quantifiers. At the endpoints of the columns are also fuzzy syllogisms which contain classical quantifiers or new forms of classical quantifiers. Fuzzy syllogisms are proved by syntactic proofs, but we also use monotonicity to prove the strong validity of fuzzy syllogisms. We use monotonicity in three ways - to strengthen the first premise, to strengthen the second premise, or to weaken the conclusion. We also use Proposition 2 for the proofs.

4.1. Figure I

In the proofs of the strongly valid syllogisms in Theorems 8, 10, 11 and 12, we use monotonicity to weaken the conclusion. In these Theorems, we can see that we can order strongly valid syllogisms by monotonicity into triangles.

In Theorem 15, we order, by monotonicity, the strongly valid syllogisms into columns. In Theorem 15, we proved the syllogisms by strengthening the first premise.

4.2. Figure II

The structures of the syllogisms of Figure II are similar to the structures of Figure I. We use monotonicity to weaken the conclusion in the proofs of Theorems 16–19. As we can see in these Theorems that we ordered the strongly valid syllogisms into triangles by monotonicity.

In Theorem 21, we ordered the syllogisms into columns by monotonicity. In the proof, we used monotonicity to strengthen the first premise.

In Theorem 22, we can find eight strongly valid syllogisms. We showed the proof of syllogism (*A)Ai-II, in which we can see that its presupposition is a formula $(\exists x)(\neg Mx)$, but the middle formula in this syllogism is (Mx). This is a consequence of the property of contraposition (Lemma A1(h)). The formula representing the presupposition is related to the assumption that all formulas are not empty.

4.3. Figure III

Figure III is different than previous figures. Firstly, on this figure, we can prove non-trivial syllogisms. Non-trivial syllogisms are syllogisms, as we said in the introduction, which contain intermediate quantifiers in both premises. This group of syllogisms is specific in that valid syllogisms work with a *common presupposition*. While in the previous figures, it was enough to always assume the presupposition and thus the non-emptiness of the fuzzy set for one fuzzy set, for non-trivial syllogisms it is necessary to assume the non-emptiness of the fuzzy set in the antecedent in both premises. This assumption is represented by the formula below.

$$(\exists x)((B|z)\, x\, \&\, (B|z')\, x). \quad (27)$$

We denote the Formula (27) as a common presupposition of existential import of two fuzzy intermediate quantifiers $(Q^{\forall}_{Ev} x)(B, A)$ and $(Q^{\forall}_{Ev} x)(B, \neg A)$.

We can order strongly valid syllogisms by monotonicity into triangles. As we can see in Theorems 24 and 26–28, these triangles are oriented differently than the triangles in Figure I and Figure II. In the proofs of Theorems 24 and 26–28, we strengthen the second premise to obtain strongly valid syllogisms in the columns. To obtain the strong validity the syllogisms in the first row, we strengthen the first premise.

In Theorem 29, we can order the strongly valid syllogisms into columns by monotonicity. In this Theorem, we use monotonicity to weaken the conclusion.

4.4. Figure IV

In Figure IV, we order strongly valid syllogisms only into columns by monotonicity. We can see that in Theorem 32, we use monotonicity to strengthen the second premise. We can also see that in Theorem 33, we use monotonicity to weaken the conclusion. Finally, we can see that in Theorem 35, we use monotonicity to strengthen the first premise.

5. Conclusion and Future Work

In the article, we followed up on previous results concerning the formal proof of fuzzy logic syllogisms in fuzzy natural logic. In the introduction to the article, we first set out the motivation for this, with various references to application areas that address the issue of fuzzy generalized quantifiers. We also introduced the reader to the main mathematical territories that shape natural fuzzy logic. The main results are contained in the third section, where we first presented the mathematical definitions of fuzzy intermediate quantifiers that form a graded Peterson's cube of opposition. We managed to formally prove, in the formal mathematical system, several new forms of logical syllogisms, the validity of which we semantically verified in the finite model. The main result is that all syntactically proven fuzzy syllogisms hold in every model.

We see further development of this article in two directions. We will first focus on extending the structure of valid fuzzy logical syllogisms by more premises. In the second part, we would like to mathematically propose Peterson's rules of distributivity, quality, and quantity for verifying the validity of logical syllogisms related to a graded Peterson's cube of opposition. The second main objective for the future is to program an algorithm based on these rules and verify the validity of new forms of fuzzy syllogisms automatically.

Author Contributions: Conceptualization: K.F. and P.M.; methodology, K.F. and P.M.; validation, K.F. and P.M.; formal analysis, K.F. and P.M.; investigation, K.F. and P.M.; resources, K.F. and P.M.; writing—original draft preparation, K.F. and P.M.; writing—review and editing, K.F. and P.M.; visualization, K.F. and P.M.; supervision, P.M.; project administration, P.M. All authors have read and agreed to the published version of the manuscript.

Funding: The work was supported by ERDF/ESF by the project "Centre for the development of Artificial Inteligence Methods for the Automotive Industry of the region", No. CZ.02.1.01/0.0/0.0/17-049/0008414.

Data Availability Statement: Not applicable.

Conflicts of Interest: The authors declare no conflict of interest.

Appendix A

Appendix A.1. Main Properties of Ł-FTT

The following properties are provable in Ł-FTT and will be used in the proofs.

Lemma A1. *(Propositional properties [32]) Let $A, B, C, D \in Form_o$. Then, the following is provable:*

(a) $\vdash ((A\&B) \Rightarrow C) \equiv (A \Rightarrow (B \Rightarrow C))$;
(b) $\vdash (A \Rightarrow (B \Rightarrow C)) \Rightarrow (B \Rightarrow (A \Rightarrow C))$;
(c) $\vdash (A\&B) \Rightarrow A; (A \wedge B) \Rightarrow A; (A\&B) \Rightarrow (A \wedge B)$;
(d) $\vdash (A\&B) \equiv (B\&A)$;
(e) $\vdash (B \Rightarrow C) \Rightarrow ((A \Rightarrow B) \Rightarrow (A \Rightarrow C))$;
(f) $\vdash (C \Rightarrow A) \Rightarrow ((C \Rightarrow B) \Rightarrow (C \Rightarrow (B \wedge A)))$;
(g) $\vdash (A \Rightarrow B) \Rightarrow ((A \wedge C) \Rightarrow (B \wedge C))$;
(h) $\vdash (A \Rightarrow B) \Rightarrow (\neg B \Rightarrow \neg A)$;
(i) $\vdash ((A \Rightarrow B)\&(C \Rightarrow D)) \Rightarrow ((A\&C) \Rightarrow (B\&D))$.

Lemma A2. *(Properties of quantifiers [32]). Let $A, B \in Form_o$ and $\alpha \in Types$. Then, the following is provable:*

(a) $\vdash (\forall x_\alpha)(A \Rightarrow B) \Rightarrow ((\forall x_\alpha)A \Rightarrow (\forall x_\alpha)B)$;
(b) $\vdash (\forall x_\alpha)(A \Rightarrow B) \Rightarrow ((\exists x_\alpha)A \Rightarrow (\exists x_\alpha)B)$;
(c) $\vdash (\forall x_\alpha)(A \Rightarrow B) \Rightarrow (A \Rightarrow (\forall x_\alpha)B)$, x_α is not free in A;
(d) $\vdash (\forall x_\alpha)(A \Rightarrow B) \Rightarrow ((\exists x_\alpha)A \Rightarrow B)$, x_α is not free in B.

Lemma A3. *Let T be a theory and $A, B, C, D \in Form_o$. If $T \vdash A \Rightarrow (B \Rightarrow C)$, then $T \vdash A \Rightarrow ((B \wedge D) \Rightarrow (C \wedge D))$.*

In the proofs, we also use the rules of modus ponens and generalization, which are derived rules in our theory.

Theorem A1 ([32]). *Let T be a theory, and $A, B \in Form_o$ and $\alpha \in Types$.*

- If $T \vdash A$ and $T \vdash A \Rightarrow B$, then $T \vdash B$;
- If $T \vdash A$, then $T \vdash (\forall x_\alpha)A$.

Appendix A.2. Graded Peterson's Square of Opposition

The characteristics and position of the above-mentioned fuzzy intermediate quantifiers were studied using a graded Peterson's square of opposition. In this part of the article, we will not deal with the whole construction of the square in detail (for details, see [19]). We will recall the main definitions that form the before-mentioned square of oppositions, and show the connection between the property of monotonicity and the property of sub-altern.

Definition A1. *Let T be a consistent theory of Ł-FTT, $\mathcal{M} \models T$ be a model, and P_1, P_2 be closed formulas.*

- P_1 and P_2 are **contraries** if $\mathcal{M}(P_1) \otimes \mathcal{M}(P_2) = 0$;
- P_1 and P_2 are **sub-contraries** if $\mathcal{M}(P_1) \oplus \mathcal{M}(P_2) = 1$;
- P_1 and P_2 are **contradictories** if both $\mathcal{M}(\Delta P_1) \otimes \mathcal{M}(\Delta P_2) = 0$ and $\mathcal{M}(\Delta P_1) \oplus \mathcal{M}(\Delta P_2) = 1$;
- P_2 is a **sub-altern** of P_1 if $\mathcal{M}(P_1) \leq \mathcal{M}(P_2)$.

The proposed mathematical definitions generalize the classical definitions that form both Aristotle's and Peterson's squares of opposition. At this point, we would like to emphasize that all formally proven syllogisms apply in every model of T^{IQ}.

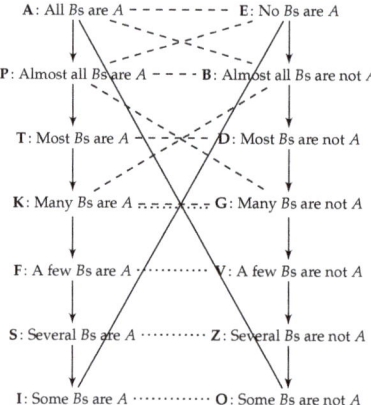

Figure A1. 7-graded Peterson's square of opposition.

Let us remind that the dashed lines denote contraries, the straight lines indicate contradictories, and the dotted lines represent subcontraries. The arrows denote the relation between superaltern–subaltern.

We continue with the theorem which represents the property of the monotonicity of quantifiers which form the 7-graded Peterson's square of opposition.

Theorem A2. *[16] Let A,...,O be intermediate quantifiers. Then, the following set of implications is provable in T^{IQ}:*

1. $T^{IQ} \vdash A \Rightarrow P, T^{IQ} \vdash P \Rightarrow T, T^{IQ} \vdash T \Rightarrow K,$
   $T^{IQ} \vdash K \Rightarrow F, T^{IQ} \vdash F \Rightarrow S, T^{IQ} \vdash S \Rightarrow I;$
2. $T^{IQ} \vdash E \Rightarrow B, T^{IQ} \vdash B \Rightarrow D, T^{IQ} \vdash D \Rightarrow G,$
   $T^{IQ} \vdash G \Rightarrow V, T^{IQ} \vdash V \Rightarrow Z, T^{IQ} \vdash Z \Rightarrow O.$

There are, of course, other studies of the graded cube of opposition. At this point, we also recall the classic cubes of opposition that were proposed by Moretti and Keyne. Gradual extensions to these two structures have been made and deeply studied by Dubois et al. in [36]. In the possibility theory, a graded extension of these cubes of opposition was analyzed by Dubois in [37].

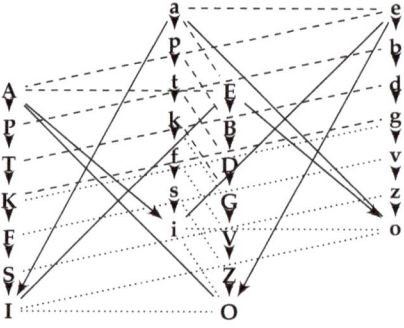

Figure A2. Graded Peterson's cube of opposition.

References

1. Kulicki, P. Aristotle's Syllogistic as a Deductive System. *Axioms* **2020**, *9*, 56. [CrossRef]
2. Castro-Manzano, J.M. Traditional Logic and Computational Thinking. *Philosophies* **2021**, *6*, 12. [CrossRef]
3. Alvarez-Fontecilla, E.; Lungenstrass, T. Aristotelian Logic Axioms in Propositional Logic. *Pouch Method Hist. Philos. Log.* **2019**, *40*, 12–21. [CrossRef]
4. Alvarez, E.; Correia, M. Syllogistic with Indefinite Terms. *Hist. Philos. Log.* **2012**, *33*, 297–306. [CrossRef]
5. Knachel, M. Categorical Syllogisms. 2019. Available online: https://www.youtube.com/watch?v=80PoPICR1DU (accessed on 2 July 2019).
6. Mostowski, A. On a Generalization of Quantifiers. *Fundam. Math.* **1957**, *44*, 12–36. [CrossRef]
7. Peterson, P. *Intermediate Quantifiers. Logic, linguistics, and Aristotelian semantics*; Ashgate: Aldershot, UK, 2000.
8. Peterson, P. On the logic of "Few", "Many" and "Most". *Notre Dame J. Form. Log.* **1979**, *20*, 155–179. [CrossRef]
9. Thompson, B.E. Syllogisms using "Few", "Many" and "Most". *Notre Dame J. Form. Log.* **1982**, *23*, 75–84. [CrossRef]
10. Zadeh, L.A. A computational approach to fuzzy quantifiers in natural languages. *Comput. Math.* **1983**, *9*, 149–184. [CrossRef]
11. Dubois, D.; Prade, H. On fuzzy syllogisms. *Comput. Intell.* **1988**, *4*, 171–179. [CrossRef]
12. Dubois, D.; Godo, L.; López de Mántras, R.; Prade, H. Qualitative Reasoning with Imprecise Probabilities. *J. Intell. Inf. Syst.* **1993**, *2*, 319–363. [CrossRef]
13. Pereira-Fariña, M.; Díaz-Hermida, F.; Bugarín, A. On the analysis of set-based fuzzy quantified reasoning using classical syllogistics. *Fuzzy Sets Syst.* **2013**, *214*, 83–94. [CrossRef]
14. Delgado, M.; Ruiz, M.; Sánchez, D.; Vila, M.A. Fuzzy quantification:a state of the art. *Fuzzy Sets Syst.* **2014**, *242*, 1–30. [CrossRef]
15. Pereira-Fariña, M.; Vidal, J.C.; Díaz-Hermida, F.; Bugarín, A. A fuzzy syllogistic reasoning schema for generalized quantifiers. *Fuzzy Sets Syst.* **2014**, *234*, 79–96. [CrossRef]
16. Novák, V. A Formal Theory of Intermediate Quantifiers. *Fuzzy Sets Syst.* **2008**, *159*, 1229–1246. [CrossRef]
17. Novák, V. A Comprehensive Theory of Trichotomous Evaluative Linguistic Expressions. *Fuzzy Sets Syst.* **2008**, *159*, 2939–2969. [CrossRef]
18. Murinová, P.; Novák, V. The structure of generalized intermediate syllogisms. *Fuzzy Sets Syst.* **2014**, *247*, 18–37. [CrossRef]
19. Murinová, P. Graded Structures of Opposition in Fuzzy Natural Logic. *Log. Universalis* **2020**, *265*, 495–522. [CrossRef]
20. Novák, V. Linguistic characterization of time series. *Fuzzy Sets Syst.* **2015**, *285*, 52–72. [CrossRef]
21. Kacprzyk, J.; Wilbik, A.; Zadrozny, S. Linguistic summarization of time series using a fuzzy quantifier driven aggregation. *Fuzzy Sets Syst.* **2008**, *159*, 1485–1499. [CrossRef]
22. Kacprzyk, J.; Zadrożny, S. Linguistic database summaries and their protoforms: Towards natural language based knowledge discovery tools. *Inf. Sci.* **2005**, *173*, 281–304. [CrossRef]
23. Yager, R. A new approach to the summarization of data. *Inf. Sci.* **1982**, *28*, 69–86. [CrossRef]
24. Yager, R. On ordered weighted averaging operators in multicriteria decision making. *IEEE Trans. Syst. Man Cybern* **1988**, *18*, 182–190. [CrossRef]
25. Yager, R. On linguistic summaries of data. In *Knowledge Discovery in Databases*; Piatetsky-Shapiro, G., Frawley, W.J., Eds.; AAAI MIT Press: Cambridge, MA, USA, 1991; pp. 347–363.
26. Yager, R. Linguistic summaries as a tool for database discovery. In Proceedings of the FUZZIEEE 1995 Workshop on Fuzzy Database Systems and Information Retrieval, Yokohama, Japan, 20–21 March 1995; pp. 79–82.
27. Kacprzyk, J.; Yager, R. Linguistic summaries of data using fuzzy logic. *Int. J. Gen. Syst.* **2001**, *30*, 133–154. [CrossRef]
28. Kacprzyk, J.; Zadrożny, S. Linguistic summarization of data sets using association rules. In Proceedings of the FUZZ-IEE'03, St. Louis, MO, USA, 25–28 May 2003; pp. 702–707.
29. Wilbik, A.; Kaymak, U. Linguistic Summarization of Processes—A research agenda. In Proceedings of the 2015 Conference of the International Fuzzy Systems Association and the European Society for Fuzzy Logic and Technology, Gijon, Spain, 30 June–3 July 2015; pp. 136–143.
30. Murinová, P.; Burda, M.; Pavliska, V. An algorithm for Intermediate Quantifiers and the Graded Square of Opposition towards Linguistic Description of Data. In *Advances in Fuzzy Logic and Technology*; Springer: Cham, Switzerland, 2017; pp. 592–603.
31. Novák, V.; Perfilieva, I.; Dvořák, A. *Insight into Fuzzy Modeling*; Wiley & Sons: Hoboken, NJ, USA, 2016.
32. Novák, V. On Fuzzy Type Theory. *Fuzzy Sets Syst.* **2005**, *149*, 235–273. [CrossRef]
33. Cignoli, R.L.O.; D'Ottaviano, I.M.L.; Mundici, D. *Algebraic Foundations of Many-Valued Reasoning*; Kluwer: Dordrecht, The Netherlands, 2000.
34. Novák, V.; Perfilieva, I.; Močkoř, J. *Mathematical Principles of Fuzzy Logic*; Kluwer: Boston, MA, USA, 1999.
35. Andrews, P. *An Introduction to Mathematical Logic and Type Theory: To Truth through Proof*; Kluwer: Dordrecht, The Netherlands, 2002.
36. Dubois, D.; Prade, H.; Rico, A. Structures of Opposition and Comparisons: Boolean and Gradual Cases. *Log. Universalis* **2020**, *14*, 115–149.
37. Dubois, D.; Prade, H.; Rico, A. Graded cubes of opposition and possibility theory with fuzzy events. *Int. J. Approx. Reason.* **2017**, *84*, 168–185. [CrossRef]

Article

A Fuzzy Grammar for Evaluating Universality and Complexity in Natural Language

Adrià Torrens-Urrutia [1,*,†], María Dolores Jiménez-López [1,†], Antoni Brosa-Rodríguez [1,†] and David Adamczyk [2,†]

- [1] Research Group on Mathematical Linguistics (GRLMC), Universitat Rovira i Virgili, 43002 Tarragona, Spain; mariadolores.jimenez@urv.cat (M.D.J.-L.); antoni.brosa@urv.cat (A.B.-R.)
- [2] Institute for Research and Applications of Fuzzy Modeling (IRAFM), University of Ostrava, 701 03 Ostrava, Czech Republic; david.adamczyk@osu.cz
- * Correspondence: adria.torrens@urv.cat
- † These authors contributed equally to this work.

Abstract: The paper focuses on linguistic complexity and language universals, which are two important and controversial issues in language research. A Fuzzy Property Grammar for determining the degree of universality and complexity of a natural language is introduced. In this task, the Fuzzy Property Grammar operated only with syntactic constraints. Fuzzy Natural Logic sets the fundamentals to express the notions of universality and complexity as evaluative expressions. The Fuzzy Property Grammar computes the constraints in terms of weights of universality and calculates relative complexity. We present a proof-of-concept in which we have generated a grammar with 42B syntactic constraints. The model classifies constraints in terms of low, medium, and high universality and complexity. Degrees of relative complexity in terms of similarity from a correlation matrix have been obtained. The results show that the architecture of a Universal Fuzzy Property Grammar is flexible, reusable, and re-trainable, and it can easily take into account new sets of languages, perfecting the degree of universality and complexity of the linguistic constraints as well as the degree of complexity between languages.

Keywords: linguistic universals; linguistic complexity; evaluative expressions; fuzzy grammar; linguistic gradience; linguistic constraints

MSC: 03B65

1. Introduction

In this paper, we propose a formal grammar for approaching the study of universality and complexity in natural languages. With this model, we afford from a mathematical point of view two key issues in theoretical linguistics. There has been a long tradition of using mathematics as a modeling tool in linguistics [1]. By formalization, we mean "the use of appropriate tools from mathematics and logic to enhance explicitness of theories" [2]. We can claim that "any theoretical framework stands to benefit from having its content formalized" [2], and complexity and language universals are not an exception.

Linguistic complexity and language universals are two important and controversial issues in language research. Complexity in language is considered a multifaceted and multidimensional research area, and for many linguists, it "is one of the currently most hotly debated notions in linguistics" [3]. On the other hand, theoretically, linguistic universals have been the subject of intense controversy throughout the history of linguistics. Their nature and existence have been questioned, and their analysis has been approached from many different perspectives.

Regarding linguistic complexity, it has been defended for a long time. The so-called dogma of equicomplexity defends that linguistic complexity is invariant, that languages are not measurable in terms of complexity and that there is no sense in trying to show that there are languages more complex than others. Given this dogma, some questions that come up are the following: if the equicomplexity axiom supports the idea that languages can differ in the complexity of subsystems, why is the global complexity of any language always identical? What mechanism slows down complexity in one domain when complexity increases in another domain? What is the factor responsable for the equi-complexity?

There has been a recent change in linguistics with regard to studies on linguistic complexity that is considerable. It has gone from denying the possibility of calculating complexity—a position advocated by most linguists during the 20th century—to a great interest in studies on linguistic complexity since 2001 [4]. During the 20th century, the dogma of equicomplexity prevailed. Faced with this position, at the beginning of the 21st century, a large group of researchers argued that it is difficult to accept that all languages are equal in their total complexity and that the complexity in one area of the language is compensated by simplicity in another. Therefore, equicomplexity is questioned, and there are monographs, articles and conferences that, in one way or another, are concerned with measuring the complexity of languages.

In fact, the number of papers published in recent years on complexity both in the field of theoretical and applied linguistics [3,5–13] highlights the interest in finding a method to calculate linguistic complexity and in trying to answer the question of whether all languages are equal in terms of complexity or, if on the contrary, they differ in their levels of complexity.

Despite the interest in studies on linguistic complexity in recent years and although, in general, it seems clear that languages exhibit different levels of complexity, it is not easy to calculate exactly these differences. Part of this difficulty may be due to the different ways of understanding the concept of complexity in the study of natural languages.

Different types of complexity can be distinguished. Pallotti [14] classifies the different meanings of the term into three types:

- *Structural complexity*, if we calculate complexity in terms of a formal property of texts related to the number of rules or patterns.
- *Cognitive complexity* is the type of complexity that calculates the cost of processing.
- *Complexity of development*. In this case, we are talking about the order in which linguistic structures emerge and are mastered in second (and possibly first) language acquisition.

These types of complexity identified by Palloti are captured by the two main types of complexity in the literature: *absolute complexity*, an objective property of the system measured in terms of the number of parts of the system, the number of interrelationships among parts, or the length of a phenomenon description [15]; and *relative complexity*, which considers language users and is related to the difficulty or cost of processing, learning or acquisition. Other common dichotomies in the literature are those that distinguish *global complexity* from *local complexity* [16] or those that establish a difference between *system complexity* and *structural complexity* [15].

To measure complexity, studies in the field propose ad hoc measures of complexity that depend on the specific interests of the analysis carried out. The proposed measures are very varied, and the formalisms used can be grouped into two types: (1) measures of absolute complexity (number of categories, number of rules, ambiguity, redundancy, etc. [16]); and (2) measures of relative complexity that face the problem of determining what type of task (learning, acquisition, processing) and what type of agent (speaker, listener, child, adult) to consider. Second language learning complexity (L2) in adults [17,18] or processing complexity [19] are examples of measures that have been proposed in terms of difficulty/cost. In many cases, other disciplines have been turned to in search of tools to calculate the complexity of languages. Information theory, with formalisms such as Shannon entropy or Kolmogorov complexity [15,16], and complex systems theory [20] are some examples of areas that have provided measures for a quantitative evaluation of linguistic complexity.

Most of the studies carried out on the complexity of natural languages adopt an absolute perspective of the concept, and there are a few that address the complexity from the user's point of view. This situation may be due to the fact that in general, it is considered that the analysis of absolute complexity presents fewer problems than that of relative complexity, since its study does not depend on any particular group of language users [16]. Relative complexity approaches compel researchers to face many problems:

- What do we mean by complex? More difficult, more costly, more problematic, more challenging?
- Different ways of using language (speaking, hearing, language acquisition, second language learning) may differ in classifying something as difficult or easy.
- When we determine that a phenomenon is complex, we must indicate for whom it is complex, since some phenomena can be very complex for one group and instead facilitate the linguistic task for other groups.
- An approach to the concept of complexity based on use requires focusing on a specific user of the language and determining the "ideal" user. How do we decide which is the main use and user of the language?

Although, as we have said, most of the works carried out adopt an absolute perspective of the concept, many specialists are interested in analyzing the relative complexity. From a relative point of view, there are three different questions that could be answered:

- From the point of view of second language learning, is it more difficult for an adult to learn some languages than to learn others?
- If we consider the processing of a language, is it more difficult to speak some languages than others?
- Focusing on the language acquisition process, does it take longer to acquire some languages than others?

Therefore, one of the possible perspectives in studies on relative complexity is one that understands complexity in terms of "learning difficulty". A relative perspective is adopted here that forces us to take the language user into account: the adult learning a language. Trudgill [17], for example, argues that "linguistic complexity equates with difficulty of learning for adults" and Kusters [18] defines complexity as "the amount of effort an outsider has to make to become acquainted with the language in question [...]. An outsider is someone who learns the language in question at a later age, and is not a native speaker". The problem that we find in these studies on relative complexity is the large number of definitions and different measures used that make the results obtained often inconsistent and not comparable. On the other hand, most complexity studies that focus on the learning process pay attention almost exclusively to the target language and the success rate of learners. In general, they do not consider the weight that the learner's mother tongue has in calculating the complexity of L2. They thus consider a kind of "ideal learner" as the basis of their analyses and focus on the complexity of the different subdomains of the target language.

In the model that we present here, we consider that in order to calculate the relative complexity of languages in terms of L2 learning, it is necessary to consider the mother tongue of the learners when calculating the relative complexity of the target language, since it seems clear that the mother tongue can facilitate or complicate the process of learning the target language and, therefore, can condition the assessment of linguistic complexity.

Regarding language universals, we can define a universal of language as a grammatical characteristic present in all or most human languages [21]. Although linguists have always been interested in discovering characteristics shared by languages, it was not until Greenberg's contribution [22], with universals based on a representative set of 30 languages, that the research topic gained popularity and depth. A decade later, despite the interest aroused by Greenberg's findings, the impossibility of improving on these results caused the study of universals to lose interest and usefulness. This object of study became relevant again a decade later, thanks to the innovations in sampling proposed by linguistic typology and authors such as Comrie [23] or Dryer [24,25]. However, this expansion of data and sampling techniques aggravates the congenital problem of linguistic universals: more and more exceptions to them appear and, therefore, the term is less reliable or representative.

In recent years, although the problem presented above has not been solved, the boost in Natural Language Processing has rescued linguistic universals from oblivion. There is a clear symbiosis between the two fields, since NLP offers many tools, resources and techniques that improve the study of universals and, above all, make it more efficient [26,27]. In turn, a true understanding of the features shared by all languages implies that recent advances in NLP, applicable only in English and a few other languages, can be more easily extended to low-resource languages.

Language universals have been investigated from two different perspectives in linguistics: on the one hand, the typological, functional or Greenbergian approach; and on the other hand, the formal or Chomskyan approach [28]. From the typological point of view, taking into account the limited data available, the universals are derived inductively from a cross-linguistic sample of grammatical structures [29]. In contrast, in the formal approach, universals are derived deductively, taking into account assumptions about innate linguistic capacity and using grammatical patterns in languages (Universal Grammar) [30].

Linguistic universals have been classified taking into account the modality and domain [21]. If we consider the *modality*, we can distinguish the following types of universals:

- *Absolute universals.* These universals are those that do not present exception and that, therefore, are fulfilled in all members of the universe. Absolute universals defend the hypothesis that a grammatical property be present in a language.
- *Probabilistic or statistical universals.* These types of universals are valid for most languages, but not for all. Probabilistic universals defend the hypothesis that a grammatical property can be present in languages with a certain degree of probability.

To the typology proposed by Moravcsik [21], we can add another common concept in the literature on universals: the concept of *rara* or *rarissima* [31,32]. In this case, we are talking about a linguistic feature that is completely opposite to the one that is considered universal. We are referring to those characteristics that are not common in languages.

Taking into account the *domain*, linguistic universals can be divided into two main types:

- *Unrestricted universals.* These type of universals may be stated for the whole universe of languages. These universals are applicable to any human language.
- *Restricted, implicational or typogical universals.* These universals affect only a part of the world's languages: those that share a given characteristic previously (if x, then y).

The above four types of universals can be schematized as follows [21]:

- Unrestricted and absolute: In all languages, Y.
- Unrestricted and probabilistic: In most languages, Y.
- Restricted and absolute: In all languages, if there is X, there is also Y.
- Restricted and probabilistic: In most languages, if there is X, there is also Y.

As Moravcsik [21] states, taking into account that it is not possible to analyze every natural language, all language universals are nothing more than mere hypotheses. As a consequence, "The empirical basis of universals research can only be (a sample of) a subset of the domain for which universals could maximally claim validity, and have traditionally been claiming validity: that of humanly possible languages. Therefore, the only viable domain for universals research, then, is all-languages-present-and-past-as-known-to-us-now" [33].

What we have said reveals both the significance of complexity and universals in language studies and the difficulties to deal with these notions. In this paper, we aim to contribute to the field by proposing a fuzzy grammar for determining the degree of universality and complexity of a natural language. By considering the degree of universality, the model calculates the relative complexity of a language. In fact, in our proposal, an inversely proportional relation between universality and complexity is established: the more universal a language is, the less complex it is. With our model, we can calculate the degree of complexity by checking the number of universal rules this language contains. The idea at the base of our model is that those languages that have high universality values will be more similar to each other, and therefore, their level of relative complexity will be lower. On the contrary, those languages with low levels of universality will have a high number of specific rules, and this will increase their level of relative complexity.

The paper is organized as follows. In Section 2, we present the models of Fuzzy Universal Property Grammar and Fuzzy Natural Logic as a strategy to define linguistic universality and language complexity as vague concepts. In Section 3, material and methods are described. In Section 4, we provide a description of the experimental results. Finally, in Section 6, we discuss the results and highlight future research directions.

2. Related Work

Both Fuzzy Property Grammars (FPGr) and Fuzzy Natural Logic (FNL) can provide strategies to approach linguistic universals and the measurement of language complexity.

Regarding linguistic universality, we consider that the concept of universal can be better defined considering a continuous scale than adopting a discrete perspective. We disregard the fact that only those linguistic rules shared by all known languages—roughly 7000—can be regarded as linguistic universals. On the other hand, we define "universality" as a continuum using the FNL's gradient features of the theory of evaluative expressions. As a result, we shall continue to respect the two extreme points that already exist: 0 (non-universal) and 1 (full-universal). However, we create a spectrum in which we shall fit those linguistic rules known as "*quasi-universals*" between these two positions. On the other hand, FPGr and FNL make it possible to devise universal models, that is language-independent models, that can be applied to all natural languages, and they use a fuzzy-gradient technique to describe linguistic universals in fuzzy terms [0, 1] rather than labeling universals with a confusing nomenclature. The times that a fuzzy universal is fulfilled or violated in a fuzzy grammar determines the fuzzy degree membership of a linguistic rule. To create a Fuzzy Universal Grammar, we will use the model of FPGr. Finally, FPGr is a cheap and reusable architecture to define linguistic phenomena and their variation and makes it possible to advance in the systematization of variation in languages.

Regarding the complexity of language, it can be captured in quantitative terms (absolute complexity), such as the more rules a grammar has, the more complex it is. Therefore, if we have a system that provides all the rules of a language, we could capture its degree of complexity under the architecture of FNL of the theory of evaluative expressions. It is also possible to measure complexity between languages with a FPGr. The more rules are shared between two languages, the less complexity will be found between those two languages. The fewer rules those languages share, the more complex they will be in relation to each other. However, this approach will have a higher cost, since it will demand checking how many rules of a targeted language are shared with respect to all the other languages. That is why we have disregarded such an approach, and we have implemented a Fuzzy Universal Property Grammar that will consider all the possible combinations of rules. Therefore, for every single set of languages, we will only need to check coincidences in our Fuzzy Universal Property Grammar. Thus, a value in terms of degree [0–1] will arise from the number of coincidences defining universality: as in what is the membership degree of a set of rules of a language with respect to a Fuzzy Universal Property Grammar. In this way, the notion of universality can help measure the relative complexity of a language, assuming that those languages that have a lot of specific rules are meant to be more complex. That is, the more universal a language is, the less complex it is; the less universal a language is, the more complex it is.

In the following, we present the models of Fuzzy Universal Property Grammar and Fuzzy Natural Logic as a strategy to define linguistic universality and language complexity as vague concepts.

2.1. Fuzzy Property Grammars for Linguistic Universality

Fuzzy Property Grammars (FPGr) [34–36] combine the formalism of Fuzzy Natural Logic [37–43] and linguistic constraints typically used in linguistics. The higher-order fuzzy logic as a formalism describes the grammar at a higher level (abstractly), enabling a mathematical formalization of the degrees of grammaticality. In comparison, linguistic constraints allow us to describe vague phenomena on a local-sentence level, characterizing the objects (constraints) as prototypical and borderline ones. Therefore, both assets are necessary to build an FPGr. There are three key concepts of an FPGr: linguistic constraint, universe of the linguistic domains and fuzzy grammar, and linguistic construction.

2.1.1. Linguistic Constraint

A linguistic **constraint** is a **relation that puts together two or more linguistic elements** such as *linguistic categories* or *parts-of-speech*. Formally, a linguistic constraint is an n-tuple $\langle A_1, ..., A_n \rangle$ where A_i are linguistic categories. We usually have $n = 2$. For example, the following linguistic categories can be distinguished for this work:

1. DET (determiner);
2. ADJ (adjectve);
3. $NOUN$ (noun);
4. $PROPN$ (proper noun);
5. $VERB$ (verb);
6. ADV (adverb);
7. $CONJ$ (conjunction);
8. $SCONJ$ (subordinate conjunction);
9. ADP (preposition).

There are four types of constraints in the Fuzzy Property Grammars (FPGr):

1. **General or universal constraints** that are valid for a universal grammar. They are built from all the possible combinations between linguistic objects and constraints.
2. **Specific constraints** that are applicable to a specific grammar.
3. **Prototypical constraints** that definitely belong to a specific grammar, i.e., their degree of membership is 1.
4. **Borderline constraints** that belong to a specific language with some degree only (we usually measure it by a number from (0, 1)).

The constraints from FPGr that we will work with to describe linguistic universality and complexity are the following (the A and B are understood as linguistic categories):

— *Linearity* of precedence order between two elements: A precedes B, in symbols $A \prec B$. For example, $DET \prec NOUN$ in "*The girl*".
— *Co-occurrence* between two elements: A requires B, in symbols $A \Rightarrow B$. For example, $ADJ \Rightarrow NOUN$ in "*the red car*".
— *Exclusion* between two elements: A and B never appear in co-occurrence in the specified construction, in symbols $A \otimes B$. That is, only A or only B occurs. For example, $PRON \otimes NOUN$ in "*He runs*".
— *Uniqueness* means that neither a category nor a group of categories (constituents) can appear more than once in a given construction. For example, there is only one $PRON$ in "*She eats pizza*".
— *Dependency*. An element A has a dependency$_i$ on an element B in symbols $A \leadsto_i B$. Typical dependencies (but not exclusively) for $_i$ are *subj* (subject), *mod* (modifier), *obj* (object), *spec* (specifier), *verb* (verb), and *conj* (conjunction).

2.1.2. Definition of a Fuzzy Property Grammar

Definition 1. *A Fuzzy Property Grammar (FPGr) is a couple*

$$FPGr = \langle U, FGr \rangle \qquad (1)$$

where U is a universe
$$U = Ph_\rho \times Mr_\mu \times X_\chi \times S_\delta \times L_\theta \times Pr_\zeta \times Ps_\kappa. \qquad (2)$$

The subscripts $\rho, \ldots, \kappa$ denote types, and the sets in Equation (2) are sets of the following constraints:

- $Ph_\rho = \{ph_\rho \mid ph_\rho \text{ is a phonological constraint}\}$ is the set of constraints that can be determined in phonology.
- $Mr_\mu = \{mr_\mu \mid mr_\mu \text{ is a morphological constraint}\}$ is the set of constraints that can be determined in morphology.
- $X_\chi = \{x_\chi \mid x_\chi \text{ is a syntactic constraint}\}$ is the set of constraints that characterize syntax.
- $S_\delta = \{s_\delta \mid s_\delta \text{ is a semantic constraint}\}$ is the set of constraints that characterize semantic phenomena.
- $L_\theta = \{l_\theta \mid l_\theta \text{ is a lexical constraint}\}$ is the set of constraints that occur on a lexical level.
- $Pr_\zeta = \{pr_\zeta \mid pr_\zeta \text{ is a pragmatic constraint}\}$ is the set of constraints that characterize pragmatics.
- $Ps_\kappa = \{ps_\kappa \mid ps_\kappa \text{ is a prosodic constraint}\}$ is the set of constraints that can be determined in prosody.

The second component is a function:
$$FGr : U \to [0,1] \qquad (3)$$

which can be obtained as a composition of functions $F_\rho : Ph_\rho \to [0,1], \ldots, F_\kappa : Ps_\kappa \to [0,1]$. Each of the latter functions characterizes the degree in which the corresponding element x belongs to each of the above linguistic domains (with respect to a specific grammar).

Technically speaking, FGr in Equation (3) is a fuzzy set with the membership function computed as follows:
$$FGr(\langle x_\rho, x_\mu, \ldots, x_\kappa \rangle) = \min\{F_\rho(x_\rho), F_\mu, \ldots, F_\kappa(x_\kappa)\} \qquad (4)$$

where $\langle x_\rho, x_\mu, \ldots, x_\kappa \rangle \in U$.

Let us now consider a set of constraints from an external linguistic input $D = \{d \mid d \text{ is a dialect constraint}\}$. Each $d \in D$ can be seen as an n-tuple $d = \langle d_\rho, d_\mu, \ldots, d_\kappa \rangle$. Then, the membership degree $FGr(d) \in [0,1]$ is a degree of grammaticality of the given utterance that can be said in arbitrary dialect (of the given grammar).

2.2. Fuzzy Property Grammars for Linguistic Universality

To take into account linguistic universality, we have to point out the following considerations to the previous definitions.

We constraint the universe of our FPGr to only the syntactic domain X_χ. At this point, it is only possible to generate all the possible constraints for the syntactic domain. However, we assume that this formulation is a proof of concept for future work on the rest of the domains.

Therefore, U-FPGr will be understood exactly as shown in Equation (2).

Definition 2. *A Universal Fuzzy Property Grammar (U-FPGr) is a couple*
$$U\text{-}FPGr = \langle U, FGr \rangle \qquad (5)$$

However, its (linguistic) universe in written language stands for a simplified version of an $FPGr$ because only the syntactical domains (x) are relevant for the proof of concept that we are presenting in this work: $< x >$, the others are neglected: $FPGr = < \overline{U}, \overline{FGr} >$.

Definition 3. *U-FPGr is*

$$\overline{U} = <x> \tag{6}$$

In this case, $\overline{U}$ is generated as a Cartesian product of all the possible constraints.

$$U = Pos_\alpha \times Dep_\beta \times X_\chi \times Pos_\alpha \times Dep_\beta \times Pos_\alpha \times Dep_\beta. \tag{7}$$

The subscripts $\alpha, \ldots, \gamma$ denote types, and the sets in Equation (7) are sets of the following constraints:

- $Pos_\alpha = \{pos_\alpha \mid pos_\alpha \text{ is a linguistic category in terms of part-of-speech}\}$ is the set of linguistic categories that can be determined in all languages.
- $Dep_\beta = \{dep_\beta \mid dep_\beta \text{ is a linguistic dependency}\}$ is the set of dependencies that can be determined in all languages.
- $X_\chi = \{x_\chi \mid x_\chi \text{ is a syntactic constraint}\}$ is the set of constraints that characterize syntax.

From the linguistic point of view, each combination of $U = Pos_\alpha \times Dep_\beta$ is interpreted as a linguistic element such as a noun with subject dependency $NOUN_{[nsubj]}$, a determiner with a determiner dependency $DET_{[det]}$, or a verb as the root of the sentence $VERB_{[root]}$. Therefore, by repeating this three times, we assume that all the rules follow a linguistic constituent, such as a linguistic element (category and dependency) in relation in terms of syntactic linguistic constraints with another element (category and dependency) and a third element (category and dependency). Because of the fact that some constraints do not need this third element, we will include in our universe the possibility of having a rule without the third element.

Any language that will be computed in terms of linguistic universality will need to follow this formalism to describe its universe. The targeted language will be our linguistic input $L = \{l \mid l \text{ is a language constraint}\}$. Each $l \in L$ can be seen as an n-tuple $l = \langle l_\alpha, l_\beta, \ldots, l_\chi \rangle$. Then, the membership degree $FGr(l) \in [0,1]$ is a degree of universality given a language as a set. As seen, this is just an adaptation of how FPGr treats grammaticality. Therefore, the universality of a targeted language is computed in terms of being grammaticality understood as the membership degree of a targeted language set with respect to U-$FPGr$. Therefore, our gradient model suggests the convenience of a terminological change. We consider that it is not necessary to define our proposal as a "search for universals" task. However, on the contrary, what we intend is to search for or define a "spectrum of the universal", or what is the same, any linguistic rule that can fit to a membership degree of universality in terms of [0, 1].

Additionally, we have implemented an $IF - THEN$ rule to assign a weight value to each rule of the U-$FPGr$.

- If a rule in L coincided with a rule in the U-$FPGr$, then add +1;
- The more weight a rule has, the more universal it is in a representative set;
- The less weight a rule has, the less universal it is in a representative set.

This is quite a natural way of representing universality, since our knowledge of the universals is dependent on the system of language that we know. A rule that might be considered as a universal can become a *quasi-universal* in the moment that new languages are discovered, and such languages do not consider such a rule. Therefore, we are always computing universality in terms of a finite representative set out of the infinite sets of languages. In this case, U-$FPGr$ is flexible and re-usable, since it can update the weight of universality according any new language inserted as a linguistic input.

2.3. Fuzzy Natural Logic Computing Universals and Linguistic Complexity with Words

In order to better grasp gradient terminology as it relates to linguistic universals and complexity, we propose to compute the continuum with natural language words. For this, the concepts of *universality* and *complexity* are assumed.

Fuzzy natural logic is based on six fundamental concepts, which are the following: the concept of *fuzzy set*, Lakoff's *universal meaning hypothesis*, the *evaluative expressions*, the concept of *possible world*, and the concepts of *intension* and *extension*. The most remarkable aspect of this work is the theory of *evaluative linguistic expressions*.

An evaluative linguistic expression is defined as an expression used by speakers when they want to refer to the characteristics of objects or their parts [37,38,40–44] such as *length, age, depth, thickness, beauty,* and *kindness*, among others. In this case, we will take into account *"universality"* and *"complexity"* as evaluative expressions.

FNL assumes that the simple evaluative linguistic expression has the general form:

$$\langle intensifier \rangle \langle TE\text{-}head \rangle \qquad (8)$$

$\langle TE\text{-}head \rangle$ can be grouped together to form a *fundamental evaluative trichotomy* consisting of two antonyms and a middle term, for example $\langle good, normal, bad \rangle$. For our work, we will take into account the trichotomy of $\langle low, medium, high \rangle$. In this sense, as proposed in [45], the membership scale of universality in linguistic rules recognize:

- *High Satisfied Universal.* Linguistic rules that trigger a *high* truth value of satisfaction in $U\text{-}FPGr$, therefore, they are found satisfied in quasi-all languages. This fuzzy set includes those rules known as *Full Universals*, absolute rules, which are located in (almost) all languages.
- *Medium Satisfied Universal.* Linguistic rules that trigger a *medium* truth value of satisfaction in $U\text{-}FPGr$; therefore, they are found satisfied in the overall average of languages.
- *Low Satisfied Universal.* Linguistic rules that trigger a *low* truth value of satisfaction in $U\text{-}FPGr$; therefore, they are found satisfied in almost none of the languages.

The value of complexity is obtained from $IF - THEN$ rules such as:

Definition 4. *We characterize fuzzy $IF - THEN$ rules for complexity as follows.*

- *IF a rule is a High Universal, THEN the value of complexity is low.*
- *IF a rule is a Medium Universal, THEN the value of complexity is medium.*
- *IF a rule is a Low Universal, THEN the value of complexity is high.*

Similarly, we can express:

- *IF the value of complexity is high, THEN the rule is a low universal.*
- *IF the value of complexity is medium, THEN the rule is medium universal.*
- *IF the value of complexity is low, THEN the rule is high universal.*

The membership scale of complexity in linguistic rules is [45]:

- *Low Complexity.* Linguistic rules that have a *high* truth value in terms of weight in $U\text{-}FPGr$. They are found satisfied in quasi-all languages. This fuzzy set includes rules known as *full universals*, absolute rules, which are located in (almost) all languages.
- *Medium Complexity.* Linguistic rules that have a *medium* truth value in terms of weight in $U\text{-}FPGr$: rules found in the overall average of languages.
- *High Complexity.* Lnguistic rules that have a *low* truth value in terms of weight in $U\text{-}FPGr$: rules satisfied in almost none of the languages.

A *possible world* is defined as a specific context in which a linguistic expression is used. In case of evaluative expressions, it is characterized by a triple $w = \langle v_L, v_S, v_R \rangle$. Without loss of generality, it can be defined by three real numbers $v_L, v_S, v_R \in \mathbb{R}$ where $v_L < v_S < v_R$.

Intension and extension: Our intension will be simply the membership degree [0–1], while our extension will be dependent on the number of languages we are taking into account in a representative set for evaluating universality and complexity.

Figure 1 represents how Fuzzy Natural Logic accounts for the fuzzy-gradient notion of universality in fuzzy sets. The fuzzy limits between sets must be established. In terms of mathematical fairness rather than from a cognitive perspective, the possible world of 7000 languages has been divided into three parts for each fuzzy set. Therefore, roughly, each set is computed by 2333 language grammars. The proposed cut-off could be changed. However, we consider that there would not be a big change between the perceived perspective of the fuzzy transitions and the three-cut part criteria. We claim that the concept of universality would be better captured with a trichotomical expression of ⟨*small* − *medium* − *big*⟩ in terms of ⟨*low* − *medium* − *high*⟩. This new way of accounting for universals may have advantages over the classical nomenclature found in the literature [29,46,47] (*universal trend, statistical universal, rara, rarisima, typological generalization*, etc.).

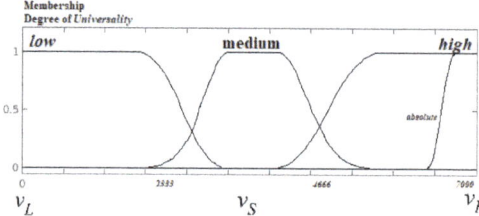

Figure 1. Linguistic Universality as a Evaluative Expression.

The advantages of the proposed model can be summarized as follows:

- The model presents a consistent classification without contradictions in terms of degree for the concepts of universality and complexity.
- The model provides a characterization of the vagueness of linguistic universality. This characterization fits the data surveys (atlas) that quatitatively collect linguistic universals terms such as WALS [48] and the Universals Archive [49].

The proposed model aims to collect the work already done in linguistics and present a universal characterization for the description of fuzzy linguistic universals and linguistic complexity.

3. Materials and Methods

3.1. A Fuzzy Universal Grammar with a Representative Set

From the 7000 languages in the world (an oscillating and debatable number), there are still a large number of them without adequate documentation. Therefore, when one wants to predict possible trends in the set of human languages as a whole, one has to investigate a selection of languages, hoping that the results will be extensible to the rest. This extension of languages is what we will consider in FNL our extension regarding the possible world of our evaluative expression of the notions of linguistic "*universality*" and "*complexity*". To this end, creating a representative and balanced set is essential. However, this task is by no means easy, as there are many other limitations [29,50,51].

For this reason, linguistic typology has classically proposed different ways of configuring a set that is as varied and independent as possible in order to be as close as possible to the reality of the 7000 languages. The selection of this independence between languages can be based on different criteria: typological, genetic, areal or a combination of them. However, it is still very difficult to find perfect samples due to what is known as bibliographical bias: the data available to us are very limited.

The representative set is build under the data from linguistic corpus. Such data allow us to create sets of languages. Working with a linguistic corpus helps us to obtain a deeper and more quantitative knowledge of cross-linguistic tendencies [52,53]. The problem with this methodology is that the available data are still very limited given its novelty and level of depth, especially in comparison with other resources based on manual notes such as the World Atlas of Linguistic Structures (WALS) [48,50]. Therefore, in order to reduce as much as possible the bibliographic bias of the languages present in Universal Dependencies [54], we have opted for a typological balance.

To create our set, we have taken into consideration three basic typological requirements that influence many other grammatical aspects in languages and their behavior [55]:

(1) Difference in the order of the subject–verb relation.
(2) Difference in the order of the object–verb relation.
(3) Difference in the order of the noun–adjective relation.

We have managed to find a good balance on points (2) and (3); however, this has not been the case with the first aspect, since it is very uncommon for the verb to precede the subject as an unmarked order. Therefore, in the representative set, its presence is also lower. We respect the proportion seen in WALS of a one-tenth part. However, it should be noted that the ascription to a particular typological order is a convenient discrete simplification [56,57].

Subsequently, we have also tried to consider the following aspects of languages to set a useful representative set:

- Languages from different genus, representatives of the main families.
- Languages from different macro-areas.
- Languages with non-dominant order in different features.
- Isolated languages.
- Agglutinating languages.
- Languages with a greater and lesser degree of ascription to other characteristics such as, for example, the use of cases.
- Corpora with enough tokens and whose source of origin does not have any type of bias, as can be the case of FAQs corpus, for example.

After setting all these requirements, primary and secondary, we decided to use the data from the Universal Dependencies corpora [58]. This data source is chosen, firstly, because it annotates a lot of different languages by part-of-speech, constituents, and dependencies, and, secondly, because it is the only formalism in which MarsaGram [59] can be applied to automatically induce sets of syntactic constraints which can be used to match coincidences between them and our *U-FPGr* . After looking at the possibilities offered by Universal Dependencies, the set established consists of the following languages:

(1) Arabic (ar.);
(2) German (de.);
(3) Basque (eu.);
(4) Spanish (es.);
(5) Estonian (et.);
(6) Indonesian (id.);
(7) Korean (ko.);
(8) Turkish (tr.);
(9) Yoruba (yo.).

Our extension will have a value of 9, and the sets of $\langle low, medium, high \rangle$ will range as follows in Figure 2:

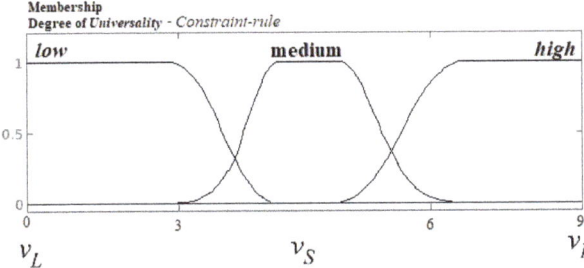

Figure 2. Linguistic Universality of the representative set as a Evaluative Expression.

1. IF a rule has between 0 and 3 coincidences, THEN the rule is a *low* universal and it is *high* in complexity.
2. IF a rule has between 4 and 6 coincidences, THEN the rule is a *medium* universal and it is *medium* in complexity.
3. IF a rule has between 7 and 9 coincidences, THEN the rule is a *high* universal and it is *low* in complexity.

Additionally, we have implemented an $IF - THEN$ rule to assign a weight value to each rule of the $U\text{-}FPGr$.

- If a rule in a set of languages coincides with a rule in the $U\text{-}FPGr$, then add +1.
- The more weight a rule has, the more universal it is in a representative set.
- The less weight a rule has, the less universal it is in a representative set.

As we have mentioned, we are aware that we cannot completely avoid bibliographical bias and, surely, the presence of a language representative of an unrepresented area or another language whose verb precedes the subject should be added. However, the model proposed here allows us to enrich the set once Universal Dependencies has such data in the future.

3.2. Application of the Tasks to Computationally Build a Universal Fuzzy Property Grammar

We have downloaded the Universal Dependency corpora for each of our sets of representative languages [58], and we have applied Marsagram [59]. Universal Dependency provides us with the constraint of dependency between constituents, and Marsagram automatically induces the constraints of *linearity*, *"co-occurrence"*, *"exclusion"*, and *"uniqueness"* over a Universal Dependency corpus.

Marsagram will provide us already with quantitative data; however, it is impossible to know which rules are coincident in a $U\text{-}FPGr$. Therefore, the interpretations that we obtain are more related with the notion of complexity rather than the notion of universality.

Marsagram presents data and rules in the following way:

Figure 3 is an extract for the marsagram from Arabic language. The rule means that verb as root excludes adjective as advcl next to ADJ as c-sub. Because of the fact that we are not interested in the other number, we will clean the data, erasing such noise. Therefore, to satisfy the coincidences, we will only keep the elements in #headproperty, symbol1, symbol2.

```
#headpropertysymbol-1symbol-2rulesoccurrencesfrequencebraking-rulesbreaking-occurrencesbreaking-frequencew0w1
VERB-rootexcludeADJ-advclADJ-csubj240.0006000.00001.00000.0006
```

Figure 3. Rule of the corpus or Arabic in Marsagram.

3.2.1. Building the Universal Fuzzy Property Grammar

To build the Universal Fuzzy Property Grammar ($U\text{-}FPGr$), we applied Equation (7). Table 1 is a representation of such a thing. To clarify, we take into account all the categories or part-of-speech (POS) for all languages according to the tagging in the universal dependencies. We then have 17 elements in POS. We consider the 64 dependencies that are present in the whole system of the universal dependencies. We consider the remaining four constraints. We combine this with two contextual linguistic elements, so, again, POS-dep is repeated twice. We obtain therefore, 4,242,536,496 rules. We repeat the same process again but considering the possibility that the rule only needs one contextual element, so POS-dep-properties-pos-dep. We obtain from that 4,734,976 rules. After summing up the both output of rules in terms of linguistic constraints, we obtain a $U\text{-}FPGr$ with 4,247,273,472 rules belonging to the syntactic domain.

Table 1. Representation of the elements involved in the production of a *U-FPGr* for syntactic constraints.

POS (17)	DEP (64)	Properties (4)	Pos-dep (1088)	4,242,538,496
ADJ	acl	exclude	empty (1088)	4,734,976
ADP	acl:relcl	precede		4,247,273,472
ADV	advcl	unicity		
AUX	advmod	require		
CCONJ	advmod:emph			
DET	advmod:lmod			
INTJ	amod			
NOUN	appos			
NUM	aux			
PART	aux:pass			
PRON	case			
PROPN	cc			
PUNCT	cc:preconj			
SCONJ	ccomp			
SYM	clf			
VERB	compound			
X	..+48			

The technical summary is the following:

(1) Read the excel file with all the grammar rules.
(2) Make all possible combinations as strings. For example, one combination can look like: "$rule_1\ rule_2\ rule_3$".
(3) Insert all possible combinations into the MongoDB database.
The MongoDB database is not necessary, but it solves many problems instead of storing universal grammars in a pure text file. The text file may be huge and have an impact on time complexity for searching for grammar rules.

3.2.2. Preparing Languages for the Universal Fuzzy Property Grammar

As mentioned in Figure 3, the data of each language set had to be cleaned and prepared before checking coincidences. Therefore, we have followed these steps:

(1) We have a set of possible grammars for n languages stored in CSV/TSV files.
(2) For one language:
  (a) Load all possible files with language grammar rules.
  (b) Preprocessing (for example: remove empty spaces, replace *, ...).

Finally, we have applied the weights to each rule, so we can measure its universality and complexity:

- For evaluating universality in one rule:
  (1) Send query to the MongoDB database for search rule in Universal Grammar.
  (2) If we found a rule in Universal Grammar, we insert a new row to the Pandas Dataframe, where the Universal Grammar column will have a current rule, and the column for the current language will have 1 and other languages will have 0.
  (3) If we not found a rule in Universal Grammar, we insert a Universal Grammar rule and 0 for all language columns.
  (4) Then, we can compute the total numbers for one rule, and we can put them into the Total column.

Table 2 is a visualization of the output, in this case, of the verb with the dependency of conjunction excluding two elements next to each other. We can observe how the final weights vary from one rule to the other. It is also clear how robust and flexible the system is. We only would need to add another column for any other language set that we would like to include to make our final weight of universality more representative.

Table 2. Example of output of rules with universality weight.

VERB-conj exclude CCONJ-cc VERB-ccomp, 0, 1, 1, 1, 1, 1, 1, 1, 0, 7, 0.7777777777777778
VERB-conj exclude CCONJ-cc PART-advmod, 0, 1, 0, 0, 1, 1, 0, 0, 0, 3, 0.3333333333333333
VERB-conj exclude CCONJ-cc ADJ-amod, 0, 0, 1, 0, 0, 1, 1, 1, 0, 4, 0.4444444444444444
VERB-conj exclude CCONJ-cc NOUN-nsubj, 0, 1, 0, 0, 0, 1, 0, 0, 0, 2, 0.2222222222222222
VERB-conj exclude CCONJ-cc PROPN-obj, 0, 1, 1, 1, 0, 1, 1, 0, 0, 5, 0.5555555555555556
VERB-conj exclude CCONJ-cc PROPN-obl, 0, 1, 1, 1, 1, 1, 0, 1, 0, 6, 0.6666666666666666
VERB-conj exclude CCONJ-cc VERB-xcomp, 0, 1, 1, 1, 1, 1, 0, 0, 0, 5, 0.5555555555555556
VERB-conj exclude NOUN-obj NOUN-obl, 0, 0, 0, 0, 0, 1, 0, 0, 0, 1, 0.1111111111111111
VERB-conj exclude NOUN-obj PROPN-obj, 0, 1, 1, 1, 0, 1, 1, 0, 0, 5, 0.5555555555555556
VERB-conj exclude NOUN-obj PROPN-obl, 0, 1, 1, 1, 1, 1, 0, 1, 0, 6, 0.6666666666666666
VERB-conj exclude NOUN-obj VERB-xcomp, 0, 1, 1, 1, 1, 1, 0, 0, 0, 5, 0.5555555555555556
VERB-conj exclude NOUN-obj ADJ-amod, 0, 0, 1, 0, 0, 1, 1, 1, 0, 4, 0.4444444444444444
VERB-conj exclude NOUN-obj VERB-ccomp, 0, 1, 1, 1, 1, 1, 1, 1, 1, 8, 0.8888888888888888
VERB-conj exclude NOUN-obj NOUN-nsubj:pass, 0, 1, 1, 0, 0, 1, 0, 0, 0, 3, 0.3333333333333333
VERB-conj exclude NOUN-obj NOUN-compound, 0, 0, 0, 0, 1, 1, 1, 1, 0, 4, 0.4444444444444444
VERB-conj exclude NOUN-obj ADP-case, 0, 0, 1, 0, 0, 1, 1, 1, 0, 4, 0.4444444444444444
VERB-conj exclude NOUN-obj PRON-obj, 0, 1, 1, 1, 1, 1, 1, 1, 1, 8, 0.8888888888888888
VERB-conj exclude NOUN-obj VERB-acl, 0, 1, 0, 0, 0, 1, 1, 0, 0, 3, 0.3333333333333333
VERB-conj exclude NOUN-obj PRON-obl, 0, 1, 1, 1, 0, 1, 0, 1, 1, 6, 0.6666666666666666
VERB-conj exclude NOUN-obj VERB-fixed, 0, 0, 0, 0, 0, 1, 0, 0, 0, 1, 0.1111111111111111
VERB-conj exclude NOUN-obj NUM-nummod, 0, 0, 0, 0, 0, 1, 1, 0, 0, 2, 0.2222222222222222
VERB-conj exclude NOUN-obj PROPN-nsubj, 0, 1, 1, 1, 1, 1, 1, 0, 0, 6, 0.6666666666666666
VERB-conj exclude NOUN-obj PART-advmod, 0, 1, 0, 0, 1, 1, 0, 0, 0, 3, 0.3333333333333333
VERB-conj exclude NOUN-obj ADJ-xcomp, 0, 1, 1, 1, 0, 1, 0, 0, 0, 4, 0.4444444444444444
VERB-conj exclude NOUN-obj PRON-nsubj, 0, 1, 1, 1, 1, 1, 1, 1, 0, 7, 0.7777777777777778
VERB-conj exclude NOUN-obj VERB-advcl, 0, 1,1, 1, 1, 1, 0, 1, 0, 6, 0.6666666666666666

To evaluate complexity, it is necessary to just negate the value (apply -1), since universality and complexity work as opposites in terms of $\langle low, medium, high \rangle$. Therefore, a rule is that if its universality weights 0.7, its complexity would be 0.3; if its universality is 0.4, its complexity would be 0.6, and so on.

4. Results

Our results can be found in four outputs: the quantitative data of Marsagram for linguistic complexity, the quantitative data regarding number of rules by weights, distribution of the weighted rules per language, and coincidences between languages.

4.1. Quantitative Data of Marsagram

Table 3 and Figure 4 show the results of the data induction from the application of Marsagram on the representative set of the universal dependency corpora. We see a high disparity between sets, among which German is the set with the most structures (32 K) and constraints (54 K), and Yoruba is the set with the fewest structures (243) and constraints (1647). We appreciate how the induction creates constraints per structure, providing a bias in evaluating complexity. That is because German is the language that has more structures. Therefore, it is the language that will have more properties.

Table 3. Quantitative data of Marsagram.

Language	Trees	Structures	Constraints	Order	s/Constraints
German	150,921	32,242	54,410	sv-ndo	1.6875504
Korean	27,363	8853	48,097	sv-ov	5.43284762
Turkish	18,687	5275	30,937	sv-ov	5.86483412
Spanish	16,013	8114	28,808	ndo-vo	3.5504067
Estonian	30,972	9468	28,570	sv-vo	3.01753274
Arabic	19,738	11,226	21,062	vs-vo	1.8761803
Euskera	8993	3283	14,703	sv-ov	4.47852574
Indonesian	5593	3143	12,530	sv-vo	3.98663697
Yoruba	318	243	1647		6.77777778

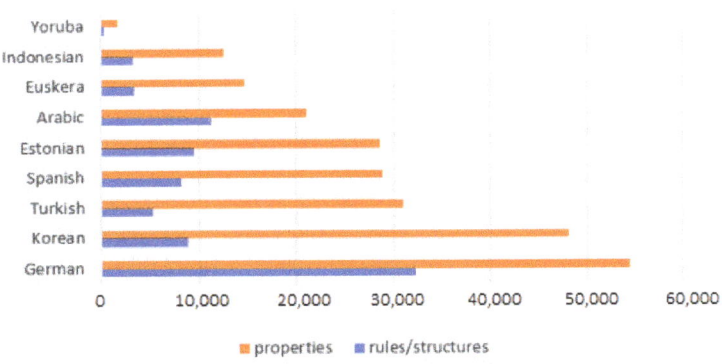

Figure 4. Quantitative data of Marsagram.

On the other hand, even though Estonian and Arabic have a similar amount of structures (9468 and 11,226) and constraints (28,570 and 21,062), they have a considerable difference in the number of trees as the input data (30,972, and 19,738). Additionally, they display fewer constraints than languages with fewer structures in the corpora, such as Turkish (5275 structures, 30,937 constraints) or Spanish (8114 structures, 28,808 constraints). Because of this data, we imply that it will be enough for future tasks to experiment with a corpus with around 20,000 dependency trees as the input data to generate structures and constraints of a language set.

However, we can apply an operation to compute the average of constraints per structure (s/Constraints column). Therefore, we will determine that languages with more constraints per structure are the most complex. According to this computation, we find that Turkish, Korean, and Euskera are the most complex language sets according to our operation, with 5.86, 5.43, and 4.47 constraints per structure, respectively. On the other hand, we disregard German and Yoruba for this first evaluation of linguistic complexity with Marsagram because of the over-generation of structures and constraints in the language set of German and the lack of data for the language set of Yoruba.

To sum up, the inductive data extracted from Marsagram can be helpful for a first glance to take into account complexity. However, it tells us nothing regarding universality, and it seems we cannot extract feasible results from asymmetric data.

4.2. Quantitative Data Regarding Number of Rules by Weights

Figure 5 shows the distribution of the number of constraints per weight. We observe that it is less frequent to find high universal constraints, which are present in all the languages. On the contrary, there seems to be a peak in constraints of weights 1 and 2. Such a thing means that two languages bear a lot of specific constraints.

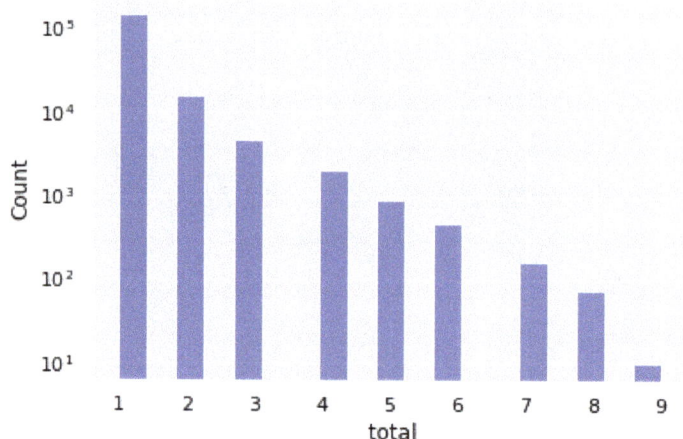

Figure 5. Quantitative data regarding number of rules by weights.

4.3. Distribution of the Weighted Rules per Set of Language

Figure 6 clarifies the data in Figure 5. The plot tends to converge in 9, since a weight of 9 means that all the sets have the constraints that weight 9. This plot displays high membership in the less universal rules. We now acknowledge that Korean, German, Turkish, and Euskera have most of the rules with weight 1 or 2. That means, in principle, that they will be the most complex languages and the ones that bear less universal constraints. It stands out how Korean has a membership degree of 1.0 for the rules of weight 1. This means that Korean has *all* the specific constraints found in the *U-FPGr*.

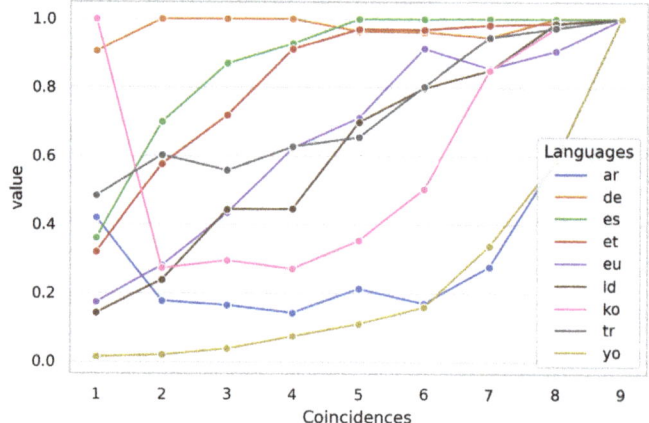

Figure 6. Distribution of the weighted rules per set of language.

4.4. Coincidences between Languages

Figure 7 displays the total number of rules found as coincident in our *U-FPGr*, and, simultaneously, the number of coincident rules between sets. For example, Spanish (es), and German (de) share 6968 rules. The matrix diagonal displays the total number of rules per set. For example, German has a total of (49,895), while Estonian has (22,477) rules. This matrix demonstrates how our system cleaned the data from Marsagram from Figure 4, where according to the induction, German had 54,510 constraints, and Estonian had 28,570. That means that Marsagram was over-generating, as we were assuming previously. This result demonstrates that our architecture is good for cleaning data from inductive algorithms that might over-generate outputs.

Figure 8 is a failed output because the data are not symmetric, so we cannot evaluate the similarity of the sets: that is, how many constraints or rules they share in terms of degree [0–1]. Additionally, this output can be read as percentages. For example, it is better not to convert these outputs into percentages. For example, AR-DE 0.020263 or AR-ES 0.0290902 correspond to 2.0263% and 2.90902%, respectively. Because the values are too small, we do not think this provides much information. It makes sense that the coincidences are so low because every set represents different languages. If the number of coincidences was higher, those two sets with high coincidences could be considered part of the same language. Normalizing the matrix was a bad option since some sets had much lower data, and it was not possible to add data. Therefore, our solution was to compute a correlation between every pair of languages to obtain a correlation matrix such as Figure 9. To make it readable, we applied a heat scale of color.

Figure 9 is the correlation matrix of our set of languages with respect to each other. This matrix gives us information about how complex the sets between each other are in terms of sharing syntactic constraints:

- Red for low quantity of shared rules;
- Yellow for quite average quantity of shared rules;
- Green for quite high quantity of shared rules;
- Blue for high quantity of shared rules.

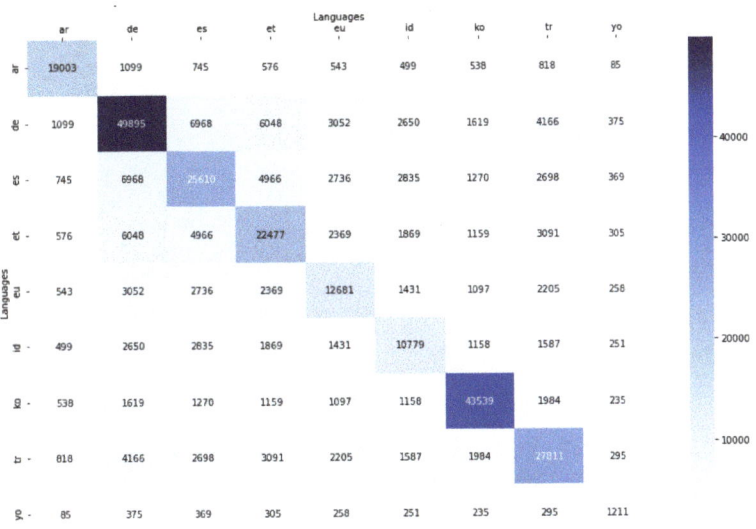

Figure 7. Coincidences between languages: number of rules.

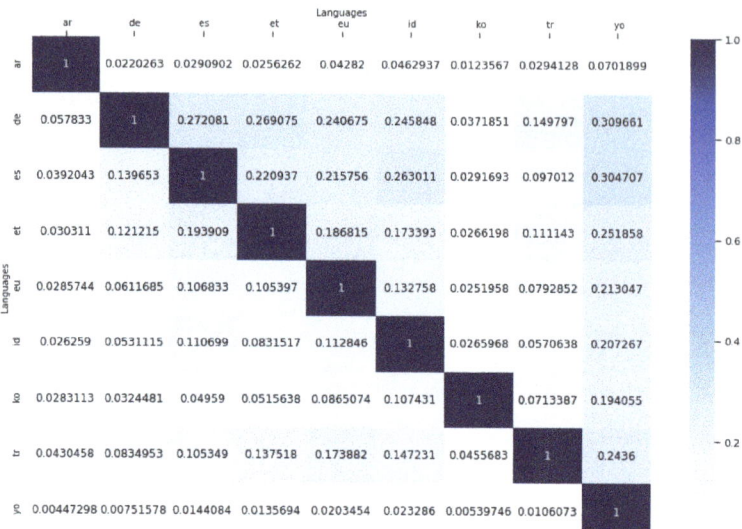

Figure 8. Coincidences between languages: degree of similarity.

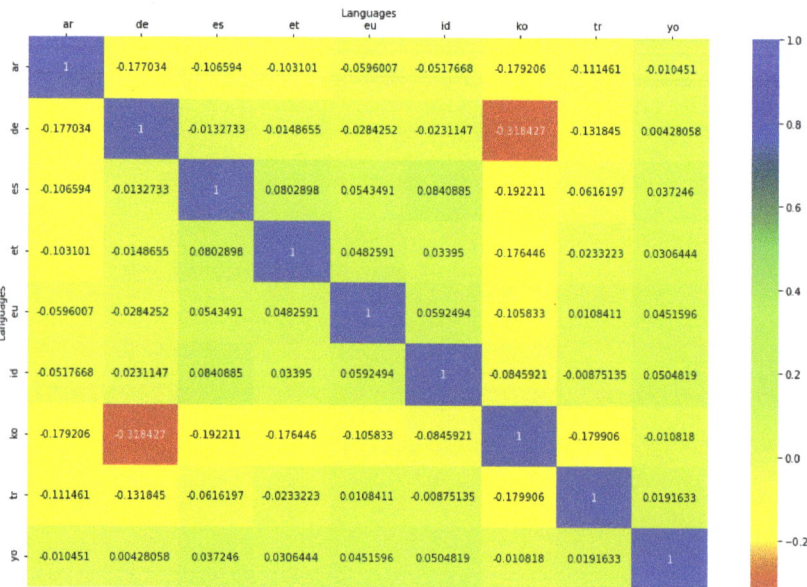

Figure 9. Coincidences between languages in a correlation matrix.

Therefore, we can represent gradually linguistic complexity in terms of evaluative expressions.

In this matrix, the closer to 1, the more similar. For example, Korean and German (ko-de) have a value of −0.3, while Korean and Spanish (ko-es) have a value of −0.19. The second relation expresses more similarity/sharing of rules than the first one since Therefore, Ko-es are more similar than ko-de and less complex because they share more rules that have a closer value to 1.

5. Discussion

With our research, we have corroborated the hypothesis that it is possible to build a system that characterizes both universality and relative complexity. The model proposed has been tested only with a proof-of-concept. However, we see clear potential despite the fact of its incompleteness. Some of the theoretical criticism that this model might receive include the following.

The first criticism can be that it only takes into account syntactic constraints. Therefore, it does not really measure a language's linguistic universality or complexity.

However, the U-$FPGr$ is a model based on the FPGr. FPGr models vagueness from any linguistic concept in terms of degrees that can be described with linguistic constraints. FPGr is compatible with the theory of evaluative expressions of FNL. Consequently, any vague linguistic concept modeled with FPGr can also be modeled with an evaluative expression with the formalism of FNL. The main issue for such a thing is that, in fact, it is always necessary that constraints characterize the concept that will be defined. Therefore, one of the improvements of both FPGr and U-$FPGr$ is that it needs a tradition that will describe each linguistic domain in terms of constraints. Such a thing is not that difficult in other domains, such as phonetics and phonology. To evaluate the universality of phones and phonemes, it would only be necessary to apply the same architecture to all phonetic and phonological tables of all the languages and/or dialects worldwide. The most coincident ones will be the most universal ones and the least complex ones. Therefore, it is not the architecture of either FPGr or U-$FPGr$ which fails to represent universality and complexity. It is the limitations of the lack of tradition in defining linguistic domains in terms of constraints.

A second criticism is that it only considers a representative set of nine languages. Therefore, the results are not a reliable definition of what syntactic constraints are universal or complex.

We claim that as a proof-of-concept, our task succeeds in showing how the U-$FPGr$ can characterize constraints and its complexity in terms of weights of universality, such as in Figure 5. From which image, it is easy to identify the sets of $\langle low, medium, high \rangle$ with 1–3, 4–6, and 7–9, respectively. Additionally, our task reveals that *high* universals are rare, which is something that goes in the same line as the linguistic literature. At the same time, it reveals that few languages, such as Korean or Turkish, trigger a lot of specific constraints. It should be very interesting and necessary that we would widen the data for further experiments, so we could corroborate if such languages keep their extreme specificity. It would also be exciting to include more language sets. However, as we have presented, the inclusion of more data per set or more sets of languages do not change the basic architecture of the U-$FPGr$. Our main issue is to provide a system that can, in fact, represent *"universality"*, and *"complexity"* as a continuum. We are looking forward to testing our model with more data and seeing if the architecture is still robust.

A third criticism is that the constraints are built upon linguistic corpus, which has induced constraints from real text sources such as Wikipedia, reviews, and newspapers. Therefore, Marsagram induced constraints of non-grammatical sentences. In such a sense, it does not represent the real complexity of the standard variant of the languages in the representative set.

There is no such a problem with respect to the induction of non-grammatical constraints regarding the standard variety of a language. FPGr is fuzzy because it takes into account both grammatical and borderline constraints. Suppose the algorithm induces constraints that are not canonical within a language or dialectal constraints; that is more than welcome. We acknowledge a language definition similar to the definition of phoneme and sound. The phoneme /s/ is the abstract representation (in a non-academic way, the summary) of several sounds, such as (s) or (z). The distinction between (s) and (z) is, in particular, that one is voiceless and the other is voiced. However, most speakers would recognize them under the same perception. This phenomenon is represented abstractly as /s/. Following this same reasoning, for FPGr, the language is an abstract representation of all the possible performances of such language. Therefore, the English language /English/ can be represented in different dialects or sociolects such as (geordie), (scouse), (apallachian), and so on. However, they are all "summarized" in the abstract representation of what English is. Therefore, if the specific constraints of these specific grammars of English are induced by an algorithm or included ad hoc by a linguist, it is more than welcome because if that constraint exists in English, it has to be included in the set of English language.

A fourth criticism is that the values in the correlation matrix display, in general, low values; therefore, it is not a valid representation of the degree of complexity.

We believe that the correlation matrix displays low values because, in fact, they are different languages. This output reinforces the idea that the architecture of $U\text{-}FPGr$ is robust. Otherwise, if the values are too similar, we might even have to say that those sets of languages are alike and, probably, they are dialects of each other. This question brings up an interesting matter for our future work: testing $U\text{-}FPGr$ with sets of dialects, or very close languages, such as the romance languages. If the values displayed in the correlation matrix are very close to value 1, it will reinforce the idea of the reliability of this architecture.

Finally, a criticism can be that the values of the relative complexity in Figure 9 are not meant to be necessarily equal between two sets since, even though two languages share a similar amount of rules, the rules that are not shared could be potentially more complex, affecting their degree of similarity. Therefore, it fails to represent possible hypothetical cases such as, for example, that for the majority of the speakers of German, it is easier to learn English than for English speakers to learn German. In this sense, the correlation of complexity should be represented asymmetrically.

However, our model does not reflect complexity in terms of difficulty between native speakers of a language. It would be very interesting to do so. Still, it would be necessary to include many more constraints in different domains to see more accurately how the constraints interact between domains and between domains with respect to other languages. It would probably be best to include the formalism of the agent-based models.

6. Conclusions and Future Work

We believe that the work presented here is a satisfactory proof-of-concept, which opens a new research line in pursuing the evaluation and definition of language universals and complexity. There is no such thing in linguistics as a "traditional method" or "fixed way" of computing such concepts of linguistic universality and complexity. Particularly, there is no proposed method that pursues to evaluate the concepts of universality and complexity as vague terms that can be defined in terms of degree. We believe that by considering such terms as gradient and fuzzy, we will be in a better position to describe multiple natural languages in linguistics, considering their idiosyncrasies and complexities. The best framework to do so is the theory of evaluative expressions of Fuzzy Natural Logic, which sets the basis to compute vague concepts with natural language and trichotomous expressions, together with Fuzzy Property Grammars, which provide the linguistic constraints to be evaluated. Furthermore, we believe this work opens a research line to make more appealing the fact of defining languages in terms of constraints to provide explicative or white-box methodologies for the characterization of languages and their features.

Regarding the future work of this research, it is necessary to test the model with data sets of dialects, and close related languages, such as romance languages. Therefore, we can test the model's outputs with data sets that, a priori, should display a larger number of coincidences and try to establish a fuzzy–numerical boundary between language and dialect. That is, to obtain a fuzzy-value which characterizes when a dialect starts to be considered a different language in terms of membership degrees. On the other hand, testing the model with larger and symmetric data sets would be necessary to reassure its robustness. Another test that has to be run in the future is to compute linguistic complexity taking into account other linguistic domains, such as computing similarity between lexicons and phonemic charts of different languages and dialects by incorporating an Optimality Theory approach. Similarly, it would be necessary to work on comparing language sets within the constraints of the morphological domain.

Author Contributions: Conceptualization, A.T.-U., M.D.J.-L. and A.B.-R.; Formal analysis, A.T.-U. and M.D.J.-L.; Software, D.A.; Writing—original draft, A.T.-U., M.D.J.-L. and A.B.-R.; Writing—review and editing, A.T.-U., M.D.J.-L. and A.B.-R. All authors have read and agreed to the published version of the manuscript.

Funding: This paper has been supported by the project CZ.02.2.69/0.0/0.0/18_053/0017856 "Strengthening scientific capacities OU II" and by Grant PID2020-120158GB-I00 funded by Ministerio de Ciencia e Innovación. Agencia Estatal de Investigación MCIN/AEI/10.13039/501100011033.

Institutional Review Board Statement: Not applicable.

Informed Consent Statement: Not applicable.

Data Availability Statement: https://universaldependencies.org/ (accessed on 12 December 2021), Marsagram corpus can be requested to adria.torrens@urv.cat.

Acknowledgments: We also want to give special thanks to Vilém Novák, Grégoire Montcheuil and Jan Hůla for their collaboration and support during this research.

Conflicts of Interest: The authors declare no conflict of interest.

References

1. Nefdt, R.M. The Foundations of Linguistics: Mathematics, Models, and Structures. Ph.D. Thesis, University of St Andrews, St Andrews, Scotland, 2016.
2. Pullum, G.K. The central question in comparative syntactic metatheory. *Mind Lang.* **2013**, *28*, 492–521. [CrossRef]
3. Kortmann, B.; Szmrecsanyi, B. *Linguistic Complexity: Second Language Acquisition, Indigenization, Contact*; Walter de Gruyter & Co.: Berlin, Germany, 2012.
4. McWhorter, J.H. The Worlds Simplest Grammars Are Creole Grammars. *Linguist. Typology* **2001**, *5*, 125–166. [CrossRef]
5. Baechler, R.; Seiler, G. *Complexity, Isolation, and Variation*; de Gruyter, Walter GmbH & Co.: Berlin, Germany, 2016; Volume 57.
6. Baerman, M.; Brown, D.; Corbett, G.G. *Understanding and Measuring Morphological Complexity*; Oxford University Press: New York, NY, USA, 2015.
7. Coloma, G. *La Complejidad de Los Idiomas*; Peter Lang Limited, International Academic Publishers: Bern, Switzerland, 2017.
8. Conti Jiménez, C. *Complejidad lingüística: Orígenes y Revisión Crítica del Concepto de Lengua Compleja*; Peter Lang Limited, International Academic Publishers: Bern, Switzerland, 2018.
9. Di Domenico, E. *Syntactic Complexity From a Language Acquisition Perspective*; Cambridge Scholars Publishing: Newcastle upon Tyne, UK, 2017.
10. La Mantia, F.; Licata, I.; Perconti, P. *Language in Complexity: The Emerging Meaning*; Springer: Berlin/Heidelberg, Germany, 2016.
11. McWhorter, J.H. *Linguistic Simplicity and Complexity: Why Do Languages Undress?* Walter de Gruyter: Berlin, Germany, 2011; Volume 1.
12. Newmeyer, F.J.; Preston, L.B. *Measuring Grammatical Complexity*; Oxford University Press: New York, NY, USA, 2014.
13. Ortega, L.; Han, Z. *Complexity Theory and Language Development: In Celebration of Diane Larsen-Freeman*; John Benjamins Publishing Company: Amsterdam, The Netherlands, 2017; Volume 48.
14. Pallotti, G. A simple view of linguistic complexity. *Second. Lang. Res.* **2015**, *31*, 117–134. [CrossRef]
15. Dahl, Ö. *The Growth and Maintenance of Linguistic Complexity*; John Benjamins Publishing Company: Amsterdam, The Netherlands, 2004; Volume 10.
16. Miestamo, M.; Sinnemäki, K.; Karlsson, F. (Eds.) Grammatical Complexity in a Cross-Linguistic Perspective. In *Language Complexity: Typology, Contact, Change*; John Benjamins Publishing Company: Amsterdam, The Netherlands, 2008; pp. 22–42.
17. Trudgill, P. Contact and simplification: Historical baggage and directionality in linguistic change. *Linguist. Typology* **2001**, *5*, 371–374.
18. Küsters, W. *Linguistic Complexity*; LOT: Utrecht, The Netherlands, 2003.
19. Hawkins, J.A. An efficiency theory of complexity and related phenomena. In *Language Complexity as an Evolving Variable*; Sampson, S., Gil, D., Trudgill, P., Eds.; Oxford University Press: New York, NY, USA, 2009; pp. 252–268.
20. Andrason, A. Language complexity: An insight from complexsystem theory. *Int. J. Lang. Linguist.* **2014**, *2*, 74–89.
21. Moravcsik, A. Explaining language universals. In *The Oxford Handbook of Language Typology*; Song, J., Ed.; Oxford University Press: Oxford, UK, 2010; pp. 69–89.
22. Greenberg, J. *Universals of Language*; The MIT Press: Cambridge, MA, USA, 1963.
23. Comrie, B. *Language Universals and Linguistic Typology*; The University of Chicago Press: Great Britain, UK, 1989.
24. Dryer, M. Large linguistic areas and language sampling. *Stud. Lang.* **1989**, *13*, 257–292. [CrossRef]
25. Dryer, M. The Greenbergian Word Order Correlations. *Language* **1992**, *68*, 81–138. [CrossRef]

26. O'Horan, H.; Berzak, Y.; Vulic, I.; Reichart, R.; Korhonen, A. Survey on the Use of Typological Information in Natural Language Processing. In Proceedings of the COLING 2016, the 26th International Conference on Computational Linguistics: Technical Papers, Osaka, Japan, 11–16 December 2016; Scalise, S., Magni, E., Bisetto, A., Eds.; The COLING 2016 Organizing Committee: Osaka, Japan, 2016; pp. 1297–1308..
27. Ponti, E.M.; O'Horan, H.; Berzak, Y.; Vulic, I.; Reichart, R.; Poibeau, T.; Shutova, E.; Korhonen, A. Modeling Language Variation and Universals: A Survey on Typological Linguistics for Natural Language Processing. *Comput. Linguist.* **2019**, *45*, 559–601. [CrossRef]
28. Lahiri, A.; Plank, F. Methods for Finding Language Universals in Syntax. In *Universals of Language Today*; Scalise, S., Magni, E., Bisetto, A., Eds.; Springer: Berlin/Heidelberg, Germany, 2009; pp. 145–164.
29. Bakker, D. Language Sampling. In *The Oxford Handbook of Linguistic Typology*; Song, J.J., Ed.; Oxford University Press: Oxford, UK, 2010; pp. 1–26.
30. Chomsky, N. *Aspects of the Theory of Syntax*; The MIT Press: Cambridge, MA, USA, 1965.
31. Wohlgemuth, J.; Cysouw, M. *Rara and Rarissima*; De Gruyter Mouton: Berlin, Germany, 2010.
32. Wohlgemuth, J.; Cysouw, M. *Rethinking Universals: How Rarities Affect Linguistic Theory*; De Gruyter Mouton: Berlin, Germany, 2010.
33. Lahiri, A.; Plank, F. What Linguistic Universals Can be True of. In *Universals of Language Today*; Scalise, S., Magni, E., Bisetto, A., Eds.; Springer: Berlin/Heidelberg, Germany, 2009; pp. 31–58.
34. Torrens Urrutia, A.; JiménezLópez, M.D.; Blache, P. Fuzziness and variability in natural language processing. In Proceedings of the 2017 IEEE International Conference on Fuzzy Systems (FUZZ-IEEE), Naples, Italy, 9–12 July 2017; pp. 1–6.
35. Torrens Urrutia, A. An approach to measuring complexity within the boundaries of a natural language fuzzy grammar. In Proceedings of the International Symposium on Distributed Computing and Artificial Intelligence, Toledo, Spain, 20–22 June 2018; Springer: Berlin/Heidelberg, Germany, 2018; pp. 222–230.
36. Torrens Urrutia, A. A Formal Characterization of Fuzzy Degrees of Grammaticality for Natural Language. Ph.D. Thesis, Universitat Rovira i Virgili, Tarragona, Spain, 2019.
37. Novák, V. The Concept of Linguistic Variable Revisited. In *Recent Developments in Fuzzy Logic and Fuzzy Sets*; Sugeno, M., Kacprzyk, J., Shabazova, S., Eds.; Springer: Berlin/Heidelberg, Germany, 2020; pp. 105–118.
38. Novák, V. Fuzzy Logic in Natural Language Processing. In Proceedings of the 2017 IEEE International Conference on Fuzzy Systems (FUZZ-IEEE), Naples, Italy, 9–12 July 2017.
39. Novák, V.; Perfilieva, I.; Dvořák, A. *Insight into Fuzzy Modeling*; Wiley & Sons: Hoboken, NJ, USA, 2016.
40. Novák, V. Mathematical Fuzzy Logic: From Vagueness to Commonsese Reasoning. In *Retorische Wissenschaft: Rede und Argumentation in Theorie und Praxis*; Kreuzbauer, G., Gratzl, N., Hiebl, E., Eds.; LIT-Verlag: Wien, Austria, 2008; pp. 191–223.
41. Novák, V. What is Fuzzy Natural Logic. In *Integrated Uncertainty in Knowledge Modelling and Decision Making*; Huynh, V., Inuiguchi, M., Denoeux, T., Eds.; Springer: Berlin/Heidelberg, Germany, 2015; pp. 15–18.
42. Novák, V. Fuzzy Natural Logic: Towards Mathematical Logic of Human Reasoning. In *Fuzzy Logic: Towards the Future*; Seising, R., Trillas, E., Kacprzyk, J., Eds.; Springer: Berlin/Heidelberg, Germany, 2015; pp. 137–165.
43. Novák, V. Evaluative linguistic expressions vs. fuzzy categories? *Fuzzy Sets Syst.* **2015**, *281*, 81–87. [CrossRef]
44. Novák, V. Mining information from time series in the form of sentences of natural language. *Int. J. Approx. Reason.* **2016**, *78*, 192–209. [CrossRef]
45. Urrutia, A.T.; López, M.D.J.; Brosa-Rodríguez, A. A Fuzzy Approach to Language Universals for NLP. In Proceedings of the 2021 IEEE International Conference on Fuzzy Systems (FUZZ-IEEE), Luxembourg, 11–14 July 2021; pp. 1–6.
46. Pagel, M. The History, Rate and Pattern of World Linguistic Evolution. In *The Evolutionary Emergence of Language*; Knight, M.S.K., Hurford, J., Eds.; Cambridge University Press: Cambridge, MA, USA, 2009; pp. 391–416.
47. Newymeyer, F. *Possible and Probable Languages: A Generative Perspective on Linguistic Typology*; Oxford University Press: Oxford, UK, 2007.
48. Dryer, M.; Haspelmath, M. The World Atlas of Language Structures Online. 2013. Available online: http://wals.info (accessed on 25 January 2022).
49. Plank, F.; Filimonova, E. The Universals Archive: A Brief Introduction for Prospective Users. *STUF Lang. Typol. Universals* **2000**, *53*, 109–123. [CrossRef]
50. Guzmán Naranjo, M.; Becker, L. Quantitative word order typology with UD. In Proceedings of the 17th International Workshop on Treebanks and Linguistic, Oslo, Norway, 13–14 December 2018; Linkoping University Electronic Press: Linkoping, Sweden, 2018; pp. 91–104.
51. Choi, H.; Guillaume, B.; Fort, K. Corpus-based language universals analysis using Universal Dependencies. In Proceedings of the Second Workshop on Quantitative Syntax (Quasy, SyntaxFest 2021), Sofia, Bulgaria, December 2021; pp. 33–44.
52. Schnell, S.; Schiborr, N.N. Crosslinguistic Corpus Studies in Linguistic Typology. *Annu. Rev. Linguist.* **2022**, *8*, 171–191. [CrossRef]
53. Levshina, N. Corpus-based typology: Applications, challenges and some solutions. *Linguist. Typology* **2021**, *26*, 129–160.
54. Nivre, J.; de Marneffe, M.; Ginter, F.; Goldberg, Y.; Hajic, J.; Manning, C.D.; McDonald, R.; Petrov, S.; Pyysalo, S.; Silveira, N.; et al. Universal dependencies v1: A multilingual treebank collection. In Proceedings of the Language Resources and Evaluation Conference, Portoroz, Slovenia, 23–28 May 2016; pp. 1659–1666.
55. Siewierska, A. *Word Order Rules*; Croom Helm: New York, NY, USA, 1988.

56. Gerdes, K.; Kahane, S.; Chen, X. Typometrics: From Implicational to Quantitative Universals in Word Order Typology. *Glossa J. Gen. Linguist.* **2021**, *6*, 17.
57. Levshina, N.; Namboodiripad, S.; Allassonnière-Tang, M.; Kramer, M.A.; Talamo, L.; Verkerk, A.; Wilmoth, S.; Garrido Rodriguez, G.; Gupton, T.; Kidd, E.; et al. Why we need a gradient approach to word order. *PsyArXiv* **2021**, 1–53. *preprint*.
58. Universal Dependency Corpora. Available online: https://universaldependencies.org/ (accessed on 12 December 2021).
59. Blache, P.; Rauzy, S.; Montcheuil, G. MarsaGram: An excursion in the forests of parsing trees. In Proceedings of the Language Resources and Evaluation Conference, Portoroz, Slovenia, 23–28 May 2016; p. 7.

Article

Describing Linguistic Vagueness of Evaluative Expressions Using Fuzzy Natural Logic and Linguistic Constraints

Adrià Torrens-Urrutia [1,*,†], Vilém Novák [2,†] and María Dolores Jiménez-López [1,†]

1. Research Group on Mathematical Linguistics (GRLMC), Universitat Rovira i Virgili, 43002 Tarragona, Spain; mariadolores.jimenez@urv.cat
2. Institute for Research and Applications of Fuzzy Modeling (IRAFM), University of Ostrava, 701 03 Ostrava, Czech Republic; vilem.novak@osu.cz
* Correspondence: adria.torrens@urv.cat
† These authors contributed equally to this work.

Abstract: In recent years, the study of evaluative linguistic expressions has crossed the field of theoretical linguistics and has aroused interest in very different research areas such as artificial intelligence, psychology or cognitive linguistics. The interest in this type of expressions may be due to its relevance in applications such as opinion mining or sentiment analysis. This paper brings together Fuzzy Natural Logic and Fuzzy Property Grammars to approach evaluative expressions. Our contribution includes the marriage of mathematical and linguistic methods for providing a formalism to deal with the linguistic vagueness of evaluative expressions by describing the syntax and semantics of these structures. We contribute to the study of evaluative linguistic expressions by proposing a formal characterization of them where the concepts of semantic prime, borderline evaluative expressions and markedness are defined and where the relation between the semantic constraints of evaluations and their sentiment can be established. A proof-of-concept of how to create a lexicon of evaluative expressions for future computational applications is presented. The results demonstrate that linguistic evaluative expressions are gradient, have sentiment, and that the evaluations work as a relation of hypernym and hyponym, the hypernym being a semantic prime. Our findings provide the basis for building an ontology of evaluative expressions for future computational applications.

Keywords: evaluative expressions; linguistic gradience; fuzzy grammar; linguistic constraints; grammaticality; sentiment analysis

MSC: 94D05

1. Introduction

This article focuses on evaluative linguistic expressions, i.e., expressions such as *small, medium, large, very deep, rather shallow, beautiful,* etc. Our main objective is to provide a formal characterization of this type of expressions by defining them in syntactic terms (determining their canonical and borderline structures) and by describing them semantically in order to account for their vagueness and characterizing their sentiment polarity. In this work, we propose a model with a single scale that allows grading and computing the semantic orientation of an evaluative expression by means of a fuzzy relationship between the semantics of these expressions and the concepts of *'positive–negative'*. With this model, we can calculate the semantic orientation of words.

Fuzzy Natural Logic (FNL) and its theory of evaluative linguistic expressions is a well-known mathematical theory [1–7]. This theory is based on Lakoff's hypothesis of universal meaning [8]. Lakoff argues that natural logic is a collection of terms and rules that come with natural language and that allow us to reason and argue in it. FNL tries to account for the basic principle by which people understand each other. This principle is universal and makes communication possible between people of different origins and cultures. An example of this universal representation of meaning is found when any speaker of any place and language and in any situation is able to understand an expression such as *"that person is far away"*. Regardless of whether the meaning of *"far"* is *"800 km"* or *"5 km"*, we understand that *"far"* means "a high value on a scale of *distance"*, with regards to a specific context. A component of FNL is also the theory of intermediate quantifiers [9–11] that are expressions such as *most, many, a lot of, a few, several*, etc..

FNL has been successfully applied in time series analysis [12–14], or data mining [15]. We claim that it can be useful for a richer extraction of semantic meaning and for more accurate sentiment analysis. The main advantage of our formalism is that it provides a unified formal analysis for the semantics of all kinds of evaluative expressions.

However, Fuzzy Natural Logic is a hardly known theory in linguistics and is barely used for linguistic semantic analysis. Most of the recent works in linguistics that involve the study of the semantics of the evaluative language is found in the area of sentiment analysis [16].

Some of the works in linguistics integrating FNL as a fundamental part of their framework can be found in the modelling of a Fuzzy Property Grammar [17–20]. FNL is essential to define the basis of a fuzzy grammar with linguistic features in such work. Furthermore, FNL describes the grammar on a higher (abstract) level. In contrast, linguistic constraints describe the grammar on a local level (at the level of the sentence). The notion of linguistic constraint stands for a relation that puts together two or more linguistic elements, allowing Fuzzy Property Grammar to model vague, highly controversial phenomena for linguists at both levels. Some of the phenomena that a Fuzzy Property Grammar models using the theory of evaluative linguistic expressions are linguistic gradience in natural language, grammaticality, linguistic complexity, and linguistic universality.

This paper provides modifications and suggestions to expand the theory of evaluative expressions towards a linguistic theory as well as future Natural Language Processing (NLP) applications. We have implemented in FNL the notions of theoretical linguistics, linguistic constraints, and characterization of borderline semantic linguistic phenomena in natural language.

Our proposal can contribute to the lexicon-based models in sentiment analysis. The issue of extracting sentiment automatically has been approached in two different ways [21,22]: by using *Machine learning techniques* or by considering *Lexicon-based approaches*. The advantage of the machine learning approach is that with a labeled data set training, a classifier can be easily built with existing classifier tools [23]. The main drawback of these techniques is that, being trained with very specific data, it is difficult to apply the model to new contexts. In contrast, lexical-based methods are robust across domains without changing dictionaries. Furthermore, applying dictionaries to a new language is a less demanding task than labeling data in a new language for a classifier [23].

Lexicon-based techniques promote a good synergy between the computational and linguistic approach since they use the linguistic information contained in the text [23]. Moreover, these approaches have been highly regarded in the recent studies.

Some techniques based on the lexicon start from the following two assumptions [23]: (1) words have a "prior polarity", having a semantic orientation independent of the context; and (2) it is possible to express the semantic orientation with a numerical value. Another important issue to consider is the fact that the semantic orientation of words is affected by the linguistic context; for this reason, it is essential to take into account contextual valence shifters.

Lexicon-based techniques provide for each word classifications of its semantic orientation. To assign each word the value of *'positive'* or *'negative'*, these classifications are based on Distribution Semantics (DSm) methods. These classifications are either discrete or gradual.

As shown in Table 1, both [24]'s SentiWordNet and [25] Macquire's Dictionary show only two classifications related to semantic sentiment: *positive* and *negative*. On the contrary, the Subjectivity Dictionary proposed by [26,27] considers in addition to the values 'positive' and 'negative' two additional elements for each of them: strong and weak. Despite a certain graduality shown by the dictionary of subjectivity, this approach still fails to capture the vague and gradient spectrum between 'positive' and 'negative'. In contrast, Ref. [23] offers a richer and more formal model in its proposed Semantic Orientation Calculator (SO-CAL) that features a graded rating on a 10-point scale, from −5 to +5. Unlike previous models, Ref. [23] makes an effort to capture both the vague and gradient nature of semantically oriented word meaning and provides a numerical value for each evaluative item that allows for more detailed extraction of sentiment and of meaning.

Table 1. Sample of semantic orientation values for lexicon-based approaches from Taboada [23].

	SentiWordNet [24] Baccianella et al. (2010)	Macquire Dictionary [25] Mohammad et al. (2009)	Subjectivity Dictionary [26] Wilson et al. (2005)	SO-CAL [23] Taboada (2016)
good	Positive	Positive	Positive (weak)	3
excellent	Positive	Positive	Positive (strong)	5
masterpiece	Positive	Positive	Positive (strong)	5
bad	Negative	Negative	Negative (strong)	−3
terrible	Negative	Negative	Negative (strong)	−5
disaster	Negative	Negative	Negative (strong)	−4

Although the models presented in Table 1 generally offer good solutions for calculating the semantic orientation of words, in this paper, we want to propose a different model that uses a single scale for grading and computing the semantic orientation of words and its sentiment together with the description of their structure in a sentence with a fuzzy grammar, i.e., describing prototypical and borderline structures in which such words appear.

The remainder of this paper is organized as follows. Section 2 introduces the basics of Fuzzy Natural Logic. Section 3 presents Fuzzy Property Grammars as a constraint-based grammar that deals with linguistic vagueness. In Section 4, our proposal of a formal characterization of evaluative expressions is presented by defining the concept and by providing a syntactic and semantic explanation of evaluative expression. Section 5 describes a proof of concept to identify trichotomic expressions in Spanish and English from SO-CAL corpus. Section 6 presents some conclusions and future research lines.

2. Fuzzy Natural Logic: Formal Prerequisites

This section summarizes the basic concepts of Fuzzy Natural Logic. It is underpinned on six key concepts:

- *Fuzzy set.* A fuzzy set A in a universe U, is a function $A : U \longrightarrow L$ where $L = [0,1]$ is a support of the standard Łukasiewicz MV-algebra $\langle [0,1], \wedge, \vee, \otimes, \rightarrow, 0, 1 \rangle$. The $\wedge, \vee$ are lattice operations (here equal to min and max), $\otimes, \rightarrow$ are Łukasiewicz conjunction and implication, respectively defined by

$$a \otimes b = \max\{0, a+b-1\}, \quad a \rightarrow b = \min\{1, 1-a+b\}$$

where $a, b \in [0,1]$. The set of all fuzzy sets on U is denoted by $\mathcal{F}(U)$. If A is a fuzzy set then its *kernel* is a set $Ker(A) = \{u \mid A(u) = 1\}$.

- *Lakoff's hypothesis of universal meaning.* According to Lakoff [8], natural logic is a collection of terms and rules that comes with natural language and allows people to reason and argue in it. This idea captures the basic principle of why people understand each other. This principle is universal in all languages and makes communication among people possible.
- *Evaluative linguistic expressions* are those expressions of natural language used by people to characterize features of objects or their parts [1,5,12] such as *length, beauty, size, complexity, economical value, kindness,* among others. The most elaborated evaluative expressions in FNL simple evaluative ones. They have the general form

$$\langle linguistic\ hedge \rangle \langle TE\text{-}adjective \rangle \quad (1)$$

  $\langle TE\text{-}adjective \rangle$ can be grouped to form a *fundamental evaluative trichotomy* consisting of two antonyms and a middle term, for example $\langle good, normal, bad \rangle$, $\langle beautiful, normal, ugly \rangle$, $\langle easy, average, complex \rangle$, etc. The triple of adjectives $\langle small, medium, big \rangle$ is taken as canonical. On the other hand, $\langle linguistic\ hedge \rangle$ makes more or less specific the meaning of the $\langle TE\text{-}adjective \rangle$. Usually it is represented by an intensifying adverb such as *"very"*, *"roughly"*, *"approximately"*, *"significantly"*, etc. This is a special expression representing a linguistic phenomenon called *hedging*. In some papers of FNL, $\langle linguistic\ hedge \rangle$ is found as $\langle linguistic\ modifier \rangle$.
- *Possible world.* It is a specific context in which a linguistic expression is used. In the case of evaluative expressions, it is characterized by a triple $w = \langle v_L, v_S, v_R \rangle$. Without loss of generality, it can be defined by three real numbers $v_L, v_S, v_R \in \mathbb{R}$ where $v_L < v_S < v_R$. These numbers represent an interval of reals $[v_L, v_S] \cup [v_S, v_R]$ where v_S is marked to emphasize position of "typically medium".
- *Intension.* The intension of a linguistic expression is a property denoted by a specific word or expression. It is independent on a concrete possible world (context) and does not change when the context is changed.
- *Extension.* The extension of a linguistic expression is the referent. It depends on the particular context and changes when the possible world changes. With the exception of few specific cases, the elements falling into an extension are delineated vaguely. In our theory, they are accompanied by degrees. For example, the expression *"very long distance"* has always a high degree in any context of distance (possible world). Thus, *"very long distance in Europe"* is *"1000 km"* when *"driving a car"*, while it is *"5 km"* when *"walking on foot"*. In such case, extensions are fuzzy sets in the intervals $[0, 1000]$ and $[0, 5]$.

Definition 1. *Let $\mathcal{A}$ be an expression of natural language. Then, its intension Int($\mathcal{A}$) is a function*

$$Int(\mathcal{A}) : W \to \mathcal{F}(V) \quad (2)$$

where W is a set of possible worlds, V is a set of some objects and $\mathcal{F}$ is a set of all fuzzy sets defined on V. The extension of $\mathcal{A}$ in a given possible world (context) $w \in W$ is a fuzzy set $Ext_w(\mathcal{A}) = Int(\mathcal{A})(w) \in \mathcal{F}(V)$.

In Figure 1 (from [6]), extensions of the expressions forming the fundamental evaluative trichotomy are depicted. Part (b) of the figure contains informal justification, how the shapes of extensions of evaluative expressions are constructed. Namely, in the given context $\langle v_L, v_S, v_R \rangle$, "small" values are those around the left bound v_L, while, e.g., "very small" or "extremely small" are part of the small ones. Similarly for "big", that are values around the right bound v_R. "Medium" values are values around the central point v_S (note that it can lay on an arbitrary place between v_L and v_R).

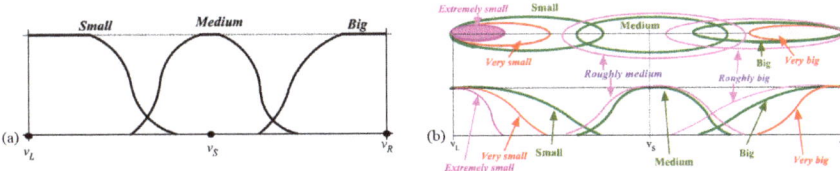

Figure 1. (a) Fuzzy sets modeling extensions of expressions that form the fundamental linguistic trichotomy; (b) Scheme of extensions of evaluative expressions with modifiers.

3. Fuzzy Property Grammar

Fuzzy Property Grammars combine the formalism of Fuzzy Natural Logic and linguistic constraints typically used in linguistics. The higher-order fuzzy logic as a formalism describes the grammar at a higher level (abstractly), enabling a mathematical formalization of the degrees of grammaticality. In comparison, linguistic constraints of a Property Grammar [28] allow us to describe vague phenomena on a local-sentence level. However, unlike a standard Property Grammar, FPGr characterizes the objects (constraints) as prototypical and borderline. Therefore, both assets are necessary to build a Fuzzy Property Grammar. In this section, the basic concepts of Fuzzy Property Grammars are presented.

A linguistic *constraint* is a relation that puts together two or more linguistic elements such as *linguistic categories*, or *parts-of-speech*. Formally, a linguistic constraint is an n-tuple $\langle A_1, ..., A_n \rangle$ where A_i are linguistic categories. We usually have $n = 2$. We consider the following linguistic categories: *DET* (determiner), *ADJ* (adjective), *NOUN* (noun), *PROPN* (proper noun), *VERB* (verb), *ADV* (adverb), *CONJ* (conjunction), *SCONJ* (subordinate conjunction) and *ADP* (preposition).

To identify linguistic categories, it is necessary to either use a part-of-speech tagger or to identify them according to the constraints of a specific language. For example, Spanish and English would have their constraints to identify a *NOUN* and an *ADJ*, which will not necessarily be shared with Chinese or Japanese. It could happen that natural languages would require an extension or a shortening of these part-of-speech. Because FPGr is closely related to the framework of Universal Dependencies (UD) [29,30], we recommend using the UD parser to identify each part of speech. UD recognises a maximum of 17 elements in a part-of-speech parsing. However, when working with FPGr in the syntactic domain, the model proposes to shorten from 17 part-of-speech to 9 (at maximum) to simplify the model and the analysis of the inputs from UD. The fact of how FPGr would identify part-of-speech with constraints UNIVERSALLY is not explored yet, since it is such a new model and mainly applied for the syntactic domain, mainly in Spanish and English. A further explanation of this can be found in Torrens–Urrutia in [19] (Chapter 7). One of the options to describe linguistic categories with constraints universally is to implement formal definitions of General Natural Syntax (GNS) from Manca and Jiménez-López [31] into FPGr.

The constraints provide a non-derivational definition of several linguistics. Linguistic constraints indicate the properties that an object (or objects) must satisfy. Thus, linguistic constraints define both the linguistic knowledge of a speaker, and the membership of a linguistic input towards a specific grammar in terms of constraint satisfaction. There are four types of constraints in Fuzzy Property Grammars:

- *General or universal constraints* that are valid for a universal grammar (any language).
- *Specific constraints* that are applicable to a specific grammar.
- *Prototypical constraints* that definitely belong to a specific grammar, i.e., their degree of membership is 1.
- *Borderline constraints* that belong to a specific language with some degree only (we usually measure it by a number from [0,1]).

The constraints that FPGr work with are the following (the *A* and *B* are understood as linguistic categories):

- *Linearity* of precedence order between two elements: A precedes B, in symbols $A \prec B$. Therefore, a violation is triggered when B precedes A. Example: *"The (DET) professor (NOUN)"*, $C_\alpha(DET \prec NOUN)$. C_α stands for satisfied constraint.
- *Co-occurrence* between two elements: A requires B, in symbols $A \Rightarrow B$. A violation is triggered if A occurs, but B does not. Example: *"The (DET) girl (NOUN) plays football"* $C_\alpha(NOUN \Rightarrow DET)$, but *"girl (NOUN) plays football"* $C_\beta(NOUN \Rightarrow DET)$. C_β stands for violated constraint.
- *Exclusion* between two elements: A and B never appear in co-occurrence in the specified construction, in symbols $A \otimes B$. That is, only A or only B occurs. Example: *"He (PRON) does yoga"*, $C_\alpha(PRON \otimes NOUN)$, but *"He (PRON) boy (NOUN) does yoga"* $C_\beta(PRON \otimes NOUN)$.
- *Uniqueness* means that neither a category nor a group of categories (constituents) can appear more than once in a given construction. For example, in a construction X, $Uniq = \{a, b, c, d\}$. A violation is triggered if one of these constituents is repeated in a construction. Example: *"The (DET) the (DET) kid that who used to be my friend"* $C_\beta($ in nominal construction: $Uniq = \{Det, Rel\})$.
- *Dependency*. An element A has a dependency$_i$ on an element B, in symbols $A \rightsquigarrow_i B$. Typical dependencies (but not exclusively) for $_i$ are *subj* (subject), *mod* (modifier), *obj* (object), *spec* (specifier), *verb* (verb), *conj* (conjunction). A violation is triggered if the specified dependency does not occur. Example: *"Andorra is a small (ADJ) country (NOUN)"*, $C_\alpha(ADJ \rightsquigarrow_{mod} NOUN)$, but *"Andorra is a small (ADJ) goodly (ADV)"*, $C_\beta (ADJ \rightsquigarrow_{mod} NOUN)$.
- *Obligation*. This property determines which elements are heads. It is expressed by the symbol □. This property tends to be avoided since FPGr pursues to define the relationship of part-of-speech without hierarchy or derivation, only at a local level. However, it is useful to define specific constructions to express the mandatory appearance of a category or linguistic feature.

Additionally, two key concepts for the formalization of Fuzzy Property Grammars are:

- *Linguistic Feature*. Features specifies when properties are going to be applied to a category. The typical feature to be represented is a linguistic function, such as the function of *subject*: $A_{[subj]}$. Features are always written and subindexed in a linguistic category, i.e., $NOUN_{[subj]}$.

    Example 1. *A property for an English grammar such as $N \prec V$ might be inaccurate since the noun can both precede and be preceded by a verb. We can specify functions and other values for a category to provide proper linguistic information thanks to the features. Therefore, property grammars can specify that a noun as a subject precedes a verb: $N_{[subj]} \prec V$. Features reinforce properties as a tool that can describe linguistic information independently of a context and more precisely represent grammatical knowledge by taking into account linguistic variation. However, features can express any type of specification from any domain. For this paper, we are interested in semantic features, which will be expressed like the following $A_{[sem:x]}$. Such characterization is useful for words that even being defined with the same category trigger different meanings, i.e., I have a snake, (snake $NOUN_{[sem:animal]}$), John is a snake, (snake $NOUN_{[sem:treacherous]}$).*

- *xCategory*. An *xCategory* is a feature which specifies that a certain category is displaying a *syntactic fit* from another category, for example, an adjective with a *syntactic fit of a noun*. All the *xCategories* are marked with a x before a prototypical category, i.e., $xADJ$. For example, in Spanish, *el boxeador (NOUN) león (NOUN)* (the lion boxer), *león (NOUN)* is performing as an adjective; therefore, *el boxeador (NOUN) león $(NOUN_{[xADJ]})$*.

Definition 2. *A Fuzzy Property Grammar (FPGr) is a couple*

$$FPGr = \langle U, FGr \rangle \quad (3)$$

where U is a universe

$$U = Ph_\rho \times Mr_\mu \times X_\chi \times S_\delta \times L_\theta \times Pr_\zeta \times Ps_\kappa. \quad (4)$$

The subscripts $\rho, \ldots, \kappa$ denote types and the sets in (4) are sets of the following constraints:

- $Ph_\rho = \{ph_\rho \mid ph_\rho$ is a phonological constraint$\}$ is the set of constraints that can be determined in phonology.
- $Mr_\mu = \{mr_\mu \mid mr_\mu$ is a morphological constraint$\}$ is the set of constraints that can be determined in morphology.
- $X_\chi = \{x_\chi \mid x_\chi$ is a syntactic constraint$\}$ is the set of constraints that characterize syntax.
- $S_\delta = \{s_\delta \mid s_\delta$ is a semantic constraint$\}$ is the set of constraints that characterize semantic phenomena.
- $L_\theta = \{l_\theta \mid l_\theta$ is a lexical constraint$\}$ is the set of constraints that occur on lexical level.
- $Pr_\zeta = \{pr_\zeta \mid pr_\zeta$ is a pragmatic constraint$\}$ is the set of constraints that characterize pragmatics.
- $Ps_\kappa = \{ps_\kappa \mid ps_\kappa$ is a prosodic constraint$\}$ is the set of constraints that can be determined in prosody.

The second component is a function

$$FGr : U \longrightarrow [0,1] \quad (5)$$

which can be obtained as a composition of functions $F_\rho : Ph_\rho \longrightarrow [0,1], \ldots, F_\kappa : Ps_\kappa \longrightarrow [0,1]$. Each of the latter functions characterizes the degree in which the corresponding element x belongs to each of the above linguistic domains (with regards to a specific grammar).

Technically speaking, $FPGr$ in (5) is a fuzzy set with the membership function computed as follows:

$$FGr(\langle x_\rho, x_\mu, \ldots, x_\kappa \rangle) = \min\{F_\rho(x_\rho), F_\mu, \ldots, F_\kappa(x_\kappa)\} \quad (6)$$

where $\langle x_\rho, x_\mu, \ldots, x_\kappa \rangle \in U$.

Let us now consider a set of constraints from an external linguistic input $D = \{d \mid d$ is a dialect constraint$\}$. Each $d \in D$ can be observed as an n-tuple $d = \langle d_\rho, d_\mu, \ldots, d_\kappa \rangle$. Then, the membership degree $FGr(d) \in [0,1]$ is a degree of grammaticality of the given utterance that can be said in arbitrary dialect (of the given grammar).

FPGr operates taking into account the notion of *linguistic construction*, originally from [32,33]. A linguistic construction is understood as a *pair of structure and meaning*.

In FPGr, linguistic constructions in written language stands for a simplified version of a FPGr because only three linguistic domains are relevant for it: the morphological domain (mr), the syntactical domain (x), the semantic domain (s), $<mr, x, s>$, the others are neglected: $FPGr = <\overline{U}, \overline{FGr}>$.

Definition 3. *A construction is:*

$$\overline{U} = <mr, x, s> \quad (7)$$

Examples of constraints in the linguistic domains in Equation (7):

- *The morphological domain $<mr>$*, which defines the part-of-speech and the constraints between lexemes and morphemes. For example, in English, the lexeme of an adjective $\prec$ (precedes) the morpheme *-ly*, and the morpheme *-ly* $\Rightarrow$ (requires) an adjective as a lexeme.

- *The syntactical domain* $<x>$, which defines the structure relations between categories in a linguistic construction or phrase. For example, in English, an adjective as a modifier $ADJ_{[mod]}$ of a noun $NOUN$ is dependent ($\rightsquigarrow$) of such a noun ($ADJ_{[mod]} \rightsquigarrow NOUN$).
- *The semantic domain* $<s>$, which defines the network-of-meanings of a language and its relation with the syntactical domain. This can be defined with semantic frames [34,35]. It is also responsible for explaining semantic phenomena as metaphorical meaning, metonymy, and semantic implausibility. For example, in English, *drinkable* object (i.e., $water_{[sem:drinkable]}$) $\Rightarrow$ (requires) *recipient* (i.e., $glass_{[sem:recipient]}$). A metonymy can be triggered with the follow $IF-THEN$ rule: If asking for something *drinkable* without *recipient* (i.e., "*may I have some water, please?*"), then *recipent* is included as a feature in the frame of *drinkable* object, i.e., $water_{[sem:drinkable, recipient]}$.

All the linguistic descriptions from now on are conducted following the framework of Fuzzy Property Grammars (FPGr). For a deeper insight of the theory, see [18–20].

4. Evaluative Linguistic Expressions with Linguistic Features

4.1. Linguistic Prerequisites

Fuzzy Natural Logic must include the following linguistic concepts in the theory of evaluative expressions. Those concepts are essential to characterize prototypical and borderline evaluative words.

- *Referent.* A referent is a person or thing to which a linguistic expression refers. For example, in the sentence *Paul talks to me*, the referent of the word *Paul* is the particular person called *Paul* who is being spoken of, while the referent of the word *me* is the person uttering the sentence. In the case of the evaluative linguistic expressions, we will understand the *referent* as the entity which *is being evaluated*.

  The referent of the evaluation can be: (1) in the linguistic input, (2) close to the evaluative word, or (3) it can be absent in the linguistic input but present in the extra-linguistic context.

  Usually, in English and Spanish language, the referent will typically be the subject of the evaluative sentence ($X_{[subj]}$) if the referent and the evaluation are in a dependency relationship with a $VERB$. Otherwise, the syntactic head of the evaluative word will establish a modifier dependency with it. This syntactic head will usually be a $NOUN$ from the evaluation. For example, in *Paul runs fast*, the referent of the evaluation *runs fast* is *Paul*, not the $VERB$ *runs*. Therefore, the fact that the evaluative word has a dependency of *modifying* a $VERB$, $ADJ \rightsquigarrow_{mod} VERB$, does not imply that it will be the referent, even though $VERB$ is a higher element in the syntactic hierarchy. Sometimes, a referent can be co-referential. That is, the same referent appears twice in a sentence. For example, *Paul had his huge backpack with him*. In such case, *Paul* and *him* refers to the same person. But, following our definition, the referent of the evaluation is not *Paul*, but the *backpack* since it is the one receiving the evaluation.

  When the referent is in the linguistic input, the evaluative word and the referent will have a relationship of dependence. Such as in *Mark watches a bad show on TV*. In this case, the referent is *a show*, and the evaluative word is *bad*. On the other hand, typically, when the evaluation and the referent are connected with a dependency through a verb, the referent will usually be the subject. For example, *Cristina thinks that Claudia is smart enough to pass her degree*. The evaluation is *is smart enough*, and the referent is *Claudia*. Both constructions have a relationship of dependence, and they are connected because of the verb. However, further exploring to set a frequent pattern of dependency between referent and evaluation is needed.
- *Linguistic pairing* stands for the task or phenomena in which linguistic features associate two linguistic elements. These can be syntactic or semantic features or both. The most interesting linguistic pairing for adding linguistic features in the evaluative linguistic expressions is the one which *pairs* the referent with the evaluative word.
- *Markedness* arises to represent the importance of the linguistic context (not extra-linguistic) for a word. A sentence α is more marked than a sentence β if α is acceptable

in less contexts than β. It can be determined either by the judgments of the speakers or by extracting the number of possible context types for a sentence [36]. Therefore, a canonical constraint or property (C_α) is never context dependent. For example, $C_\alpha(DET \prec NOUN)$. A borderline constraint or property (C_γ) is context-dependent if the degree of unacceptability triggered by its violation varies from one context to another. For example, in Spanish, canonically, determiners do not precede proper nouns, that is $C_\alpha(PROPN \otimes DET)$. However, such constraint can be violated when what the structure $\{DET, PROPN\}$ is triggering the meaning of a football team, "*El (DET) Barcelona (PROPN) era mejor con Messi que ahora*" ("*The Barcelona was better with Messi than now*"). Such a feature is important to characterize and clarify both the referent and the evaluation. Following the last sentence, the evaluation of *was better with Messi than now* would have a biased referent if the grammar would understand that *the Barcelona* meant *the city of Barcelona*, and not *the football club*.

In this case, the violation of the property, accepted only in few contexts, displays and characterizes a *marked structure* ($DET \prec PROPN_{[sem:cities]}$) (*sem* : means semantics), with a *marked meaning* ($DET \prec PROPN_{[sem:footballteam]}$), expressed as: $IFF\ C_\beta(DET \prec PROPN); AND PROPN_{[city]} THEN C_\gamma(DET \prec PROPN_{[sem:footballteam]})$.

Additionally, in semantics, structures are marked when they do not allow interchangeability of elements, and yet keep the same meaning. A non-marked structure would be: "*Mary's house, the nice one, is in Elm Street*". I change the elements and yet I can get almost same meaning "*My sister's home, the one when you were last time, is next to mine*" (knowing that *Mary* is *my sister*, *the nice one* is *the one when you were last time* and *next to mine* is *Elm Street*)'. However, this cannot happen with idioms, or other marked structures such as "*Mark is a pain in the ass*" by *Mark is a pain in the finger/eye/foot/chest/etc.*, or "*it is raining cats-and-dogs*" by *it is raining elephants-and-giraffes*, etc. Therefore, idioms tend to have marked meanings. Another example: "*No shit, Sherlock*", meaning '*judgment over a statement that was obvious*'. It is implausible to exchange such construction for other elements such as: *No turd, detective*, or *No smoke, Einstein*. Moreover, FNL should not purse to characterize extra-linguistic extreme-implied (non-)evaluations that demand extreme knowledge of the context, such as *the one when you were last time* understanding that such a construction is a very polite way of saying *I am referring to the nice one, not the ugly one*.

- *Semantic Coercion* stands for *implausibility of meaning*. Coercion is a typical effect in defining and identifying metaphors, which non-literal meanings are triggered due to the implausibility of such meaning. To summarize, it is the similar effect of markedness but focused on the semantic domain.
- *Compositionality* is typically understood as that the summarization of meanings brings us the final meaning, i.e., "*it a strong rain*", meaning: [*it + is + a + strong + rain*]. However, compositionality does not always work taking into account the previous example *it is raining cats-and-dogs*. It does not actually mean that *animals are falling from the sky!*. Therefore, the evaluative sentence *it is raining cats-and-dogs* does not fall on the logic of compositionality.
- *Lexical unit*, in FPGr, stands for those structures that work as a single part-of-speech, that is as a single unit, and, therefore, they only count as a one category. For example, in *the chief (NOUN) executive (NOUN) office (NOUN) of Amazon is Jeff Bezos*; FPGr would parse it as *the [chief executive office (NOUN)] of Amazon is Jeff Bezos*, and not as *the [chief (NOUN) executive (NOUN) office (NOUN)] of Amazon is Jeff Bezos*. Therefore, the *chief executive office* is working as a single *NOUN*. In fact, that is why the tendency is to reduce it to *CEO*, *the CEO of Amazon is Jeff Bezos*. The same applies to evaluations such as *it is raining cats-and-dogs*, which stands for *it is raining really heavily* or *it is pelting down*, or *it is bucketing down*, which stands for *it is raining very, very heavily*. Therefore, the notion of the lexical unit serves both the evaluation and its referent.

4.2. The Concept of Evaluative Expressions

An evaluation is a concept characterizing particularities of a specific object (object understood in a broad sense, as people are objects too). That is the characterization of a referent.

Examples of linguistic evaluations are the following ones:

(1) I love you very much.
   - Express affection. The verb *love* is the evaluative word. The referent is *'you'*.
(2) *Plane food* is disgusting.
   - Express a judgment. The evaluative word is *'disgusting'*. The referent is *'plane food'*. The referent operates as either a lexicalized $NOUN$ (*'plane food'* is $NOUN$), or as a $NOUN$ with an adjective fit which modifies the other $NOUN$: *'plane'* ($NOUN_{[xADJ]} \rightsquigarrow_{mod}$) *'food'* ($NOUN_{[subj]}$).
(3) *John* is not a bad person, but *he* can be a snake.
   - Express a judgment. *'Snake'* is fitting a metaphorical meaning, *'treacherous'*, and not *the animal*. The referents are *John*, and *he*.
(4) *John* looked like this (showing at a tomato) when Jennifer asked *him* for a date.
   - Express a judgment. *'this'* is a deictic work. Thus, it is empty of meaning by itself, and the context is needed to grasp the meaning that is intended to be communicated. The context reveals that it *probably* means *John was shy*. Probably is stressed since the meaning is not known for sure. Only an implication can be made by linking the linguistic input, its linguistic context, world knowledge, and communicative context.

If we accept that all the above sentences express concepts that are a matter of degrees, then we have a fuzzy set: $A : M \longrightarrow [0, 1]$. Every element can be associate with types: M_δ stands for *'John'*, M_ϵ stands for the concept of *'being bad'*, M_θ stands for *the sentiment of the concept* (positive or negative), and M_o express a membership degree, creating a complex set of functions: $A : M_\epsilon \times M_\delta \times M_\theta \to M_o$.

Therefore, an evaluative linguistic expression is an evaluative concept expressed through a language system characterizing a referent. The key elements are the evaluator and the object being evaluated.

Additionally, the linguistic evaluative expressions can be understood under the notion of *linguistic construction* [32,33], which stands that a linguistic construction is a pair of structure and meaning. Therefore, defining both linguistic syntactic and semantic features of an evaluative construction is necessary. An evaluation construction is a linguistic input evaluating a referent. This referent can be found in the linguistic input itself or can be extra-linguistic (found in the communicative context or the common knowledge).

4.3. Semantics of Evaluative Expressions

To express the generation of an evaluative linguistic expression by being the evaluator at the centre of the concept, we have provided a semantic hierarchy of the evaluation represented in Figure 2.

Figure 2 expresses the hierarchy of the different levels in which the semantics of an evaluative concept are found:

- On the first level, closer to the thought, we will find a *semantic prime*, always understanding semantic primes as those most abstract concepts which will define a fuzzy set. This idea is inspired in [37]. However, this author considers as semantic primes only the concepts of *good-bad*. While in this work, we consider primes as all those triplets that create a *fundamental trichotomous expression* such as [*ugly-medium-beautiful*], [*small- medium-big*], [*easy-medium-complex*], etc. Any prime has to be a prototypical meaning, that is, non-context dependent.

- On the second level, we would find those evaluations that are sensitive to the linguistic context. Those words that are given a specific structure trigger a non-prototypical meaning. That is the case of Example 3), in which the syntactic relation between the referent *he*, meaning a person, the verb 'to be', and the word *snake*, trigger a borderline meaning on the word snake due to the implausibility that a person would be an actual animal. This effect of implausibility is called semantic coercion [38–41]. However, it is clear that meaning **is not** a matter of probabilities in both the semantic prime, and in the linguistic context stage. The semantics of linguistic inputs related with those stages can be characterized using logic and linguistic features.
- On the third level, and further from the semantic prime, we would find those evaluations that need extra-linguistic context. Such concepts can only be understood when the communicative context is known or expresses a common knowledge shared by the interlocutors. Example 4) is a clear example of extra-linguistic meaning. Therefore, it is impossible to extract its meaning from the linguistic input's information to us. These cases are acknowledged as being *extremely borderline*. We will not formalize this type of evaluative expression here.

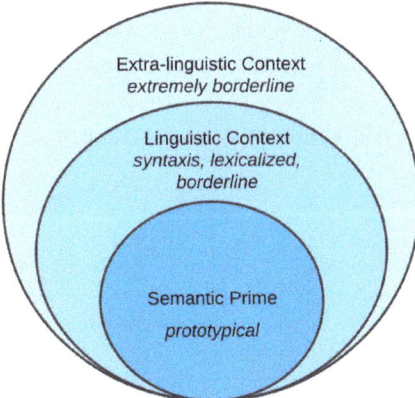

Figure 2. Semantic hierarchy of evaluative meaning

Summing up, evaluative expressions are linguistic inputs that express a concept of an evaluation that is a gradient and has a sentiment. The semantic hierarchy gives structure to identify the abstract types of the meaning of an evaluative concept. The meaning of an evaluation can be closer to a prototypical evaluation (a semantic prime), to a borderline evaluation (needing of a characterization of the linguistic context), or closer to an extremely borderline evaluation (needing of the information of the communicative context).

4.4. Syntax of Evaluative Expressions

Recall from (9) that the general form of an evaluative linguistic expression is

$$\langle linguistic\ hedge \rangle \langle TE\text{-}adjective \rangle \tag{8}$$

This form poses two problems in linguistics: the confusion of the term "hedge" and the fact that a TE-adjective is not necessarily an adjective.

The term "linguistic hedge" is used in pragmatics for what is called *hedging* defined as people use hedging when they try to avoid criticism [42–44] . Some other definitions can also be found in [45]. Hassan and Said [45] define hedges as those elements "which serve human communication as more flexible, moderate, and convincing". Hedges convey "intentional vagueness", "mitigation", "tentativeness", "politeness", "indirectness", "possibility", "evasiveness" , "lack of full commitment", etc. Hedges can be characterized as expressing the writer's attitude towards both the propositional information and his/her awareness of the readers. Words that are present in hedging strategies are *"I think"*, *"I don't know*, *"a little bit"*, *"in my view"*, *"quite"*. Some of this hedges can be found also as part of an evaluative expression; however, it is important to differentiate the intensifier strategy than the discourse hedging strategy:

- *I think* this is *quite* right, but...
- Your paint seems *a little bit* odd.

The discourse function of the pragmatics of language of the terms *"quite"* and *"a little bit"* is different from the intensifying semantic function in terms of widening or stretching the degree of vagueness of the TE-adjective *"right"*, and *"odd"*. That is why it would be better to define this first part as simply $\langle intensifier \rangle$.

Regarding the use of the concept of $\langle TE\text{-}adjective \rangle$, it is necessary to change it to $\langle TE\text{-}head \rangle$. Firstly, not necessarily an adjective will be the head of an evaluative expression itself. For example, in English, we will find differences of such things:

(1) Mark is *very intelligent*:
    $\langle intensifier\text{-} very \rangle \langle TE\text{-}head\ intelligent\ (an\ adjective) \rangle$.
(2) Mark is *an Einstein*
    $\langle intensifier\text{-} none \rangle \langle TE\text{-}head\ Einstein\ (a\ Proper\ Noun) \rangle$.
(3) I *love* Mark's intelligence *very much*
    $\langle intensifier\text{—}very\ much \rangle \langle TE\text{-}head\ love\ (a\ verb) \rangle$.

Example (1) is what we would consider a canonical evaluative expression if we consider that the prototypical evaluative heads in the English language are adjectives. That is because using an adjective as an evaluative head is very productive in English.

Example (2) has a proper noun as an evaluative head, which could be considered a borderline form of the evaluative term "intelligence" since, in English, proper nouns do not usually bear the quality of being evaluative. Additionally, we can satisfactorily apply a constituent replacement test of the word *"Einstein"* for *"intelligent"*. The test is a success since we keep the same meaning, and the constituents work similarly. Such cases will lead us to distinguish between prototypical and borderline evaluative heads. It must be kept in mind that each language will decide which linguistic categories are prototypical productive or borderline for bearing an evaluative semantic meaning. Therefore, sentences (1) and (2) are synonyms but display different evaluative heads.

Example (3) shows a verb being used as an evaluative head. This case helps us to stress why the formula of an evaluative expression in its general form should bear the notion of $\langle TE\text{-}head \rangle$ and not $\langle TE\text{-}adjective \rangle$. The evaluative verb *"love"* is not considered a borderline evaluative head here since the verb love is prototypical an evaluative verb even though verbs do not prototypically bare evaluative meanings.

We establish the following rules to determine if a word is a prototypical evaluative head or a borderline one:

- If the evaluative head is canonically and productively used as an evaluative, the evaluative head is prototypical.
- If the evaluative head is not canonically and productively used as an evaluative, but it passes a constituency test by being replaced by a prototypical evaluative head, then the evaluative head is borderline.

From the above follows that the general form of an evaluative expression should be:

$$\langle intensifier \rangle \langle TE\text{-}head \rangle \qquad (9)$$

Figure 3 represents the formal characterization in FPGr of the basic structure of a linguistic construction of an evaluative expression in terms of linguistic constraints.

Construction : [(Intensifier) □Evaluative Head]
Evaluative Expression $Int \leadsto_{xmod} EH\ EH \leadsto_{xdep} Ref_{[pairing \vee extraling]}$

Figure 3. Characterization of the universal construction of evaluative expressions.

- The elements within the construction can be found between brackets [].
- The first element is the intensifier, defined as optional, that is why it is between parenthesis ().
  - It is also defined with a linguistic dependency of modifier; meaning that the intensifier will always be dependent of the evaluative head as a modifier, no matter the language or grammar.
- The second element is the evaluative head, which is mandatory. That is why it is represented with the constraint of obligation □.
  - Additionally, the evaluative head will bare a dependency on its referent, that is, the element which is being evaluated. Because we want to keep this formalization as the most linguistically universal possible, we do not state the dependency that the evaluative bears. That is why it is represented with $\leadsto_{[xdep]}$. In English, the most common evaluative construction is found as the object of a copular verb ("to be", "feel", "seem", "become", etc.) and as the subject as its reference: Subject-Copular Verb-Evaluative Construction. In such a case, we would say that the evaluative head has a dependency on the subject.
    * It is specified that the referent Ref experience a pairing with the evaluative head, which states if they are semantically compatible or if any coercion would be involved $_{[pairing]}$. A clear case of a coerced pairing can be found in the sentence "*Mark is a giant*". The pairing between "*Mark*" and "*giant*" is implausible since "*Mark*" bares the quality of being "+ *human*" and "*giants*" does not exist, they are mythological creatures. However, the paring is possible, and a coerced meaning is interpreted. If a pairing is not present in the linguistic context, the referent would be found in the communicative context; that is, it will be extra-linguistic $_{[extraling]}$.

The possibility of having different categories depending on the language makes the notion of the evaluative head provide universality to our model.

If we use an intensifier, we can make the evaluative head more precise. FNL does not imply that adjectives are always a typical evaluative head across languages. For example, Korean and Japanese do not have adjectives in the strict sense, but can use verbs as evaluative units. These kinds of structures are also possible in English where an evaluative expression such as "*I love your food*" has a verb as its evaluative head. The meaning of this structure would be paraphrased as " *Your food is excellent*". Both sentences would be computed with a high value in the degree of judgment. Therefore, with FNL we can characterize the core/depth structure of the evaluation semantics, extracting the exact core meaning of both sentences, independently of the surface structure of the linguistic construction.

4.5. Evaluative Expressions: A Formal Characterization

A formal characterization with the FPGr of the syntactic and semantic properties that a universal construction of an evaluative expression has to fulfill is presented in Figure 4.

$$
\begin{bmatrix}
\text{Evaluative expression} \\
< insert - grammar > \\
\begin{bmatrix} \text{Intensifier}_{[INT]} \\ X_{[INT]} \rightsquigarrow_{xmod} X_{[EH]} \end{bmatrix}
\begin{bmatrix} \text{Evaluative Head}_{[EH]} \\ X_{[EH]} \Box \\ X_{[EH]} \rightsquigarrow_{[xdep]} Ref_{[pairing \vee extraling]} \\ X_{[EH]} \Rightarrow TE\langle v_L - v_S - v_R \rangle \wedge \Rightarrow LSV \begin{bmatrix} \text{PROT}_{(LSV_\alpha)} \\ \text{BORD}_{(LSV_\gamma)} \end{bmatrix} \\ \\ X_{[EH]} \Rightarrow Sentiment_{TE-Judgment\langle v_L - v_S - v_R \rangle} \begin{bmatrix} \text{PROT}_{(S_\alpha)} \\ \text{BORD}_{(S_\gamma)} \end{bmatrix} \end{bmatrix}
\end{bmatrix}
$$

Figure 4. Universal construction of evaluative expressions (syntax and semantics).

As shown in Figure 4, we consider three elements in the formal characterization of evaluative expressions:

- **The insertion of grammar**. $\langle insert - grammar \rangle$ stands for inserting an FPGr or other grammar with constraints (a universe), which will define the specific constraints of a language, as mentioned in Equation (4). This grammar will define those satisfied and violated constraints of a particular language. Therefore, FPGr will be able to characterize the degree of grammaticality of an evaluative expression with regards to a specific grammar.
- **The constraints of the intensifier**. The intensifier displays the same constraints as in Figure 3.
  - $X_{[INT]} \rightsquigarrow_{[xmod]} X_{[EH]}$ means any category X as intensifier $X_{[INT]}$ has a dependency as a type of modifier $\rightsquigarrow_{[xmod]}$ towards any category as an evaluative head $X_{[EH]}$.
- **The constraints of the evaluative head**. The evaluative head displays some of the same constraints observed, as in Figure 3, such as:
  - $X_{[EH]}\Box$ meaning any category X as an evaluative head $X_{[EH]}$ is mandatory $\Box$, leaving the intensifier as optional.
  - $X_{[EH]} \rightsquigarrow_{[xdep]} Ref_{[pairing \vee extraling]}$ means any category X as an evaluative head $X_{[EH]}$ has a non specified dependency $\rightsquigarrow_{[xdep]}$ towards a referent Ref, which can be found in the linguistic context with linguistic pairing or in the extra linguistic context $Ref_{[pairing \vee extraling]}$.
  - Additionally, the evaluative head displays semantic constraints here that are necessary to complete such a universal characterization of the constructions of an evaluative expression.
    * $X_{[EH]} \Rightarrow TE\langle v_L - v_S - v_R \rangle \wedge \Rightarrow LSV$. Meaning that any category X as an evaluative head $X_{[EH]}$ requires to be part of one of the sets of a trichotomous expression $\Rightarrow TE$, either the left one v_L, the one in the middle v_S, or the right one, v_R. And ($\wedge$), the trichotomous expression, requires to be tagged with a Linguistic Semantic Variable (LSV). The trichotomous expression can be incorporated in any constraint-based grammar in terms of hyponym and hypernymy (being the hypernym interpreted as our semantic prime) simply by using $\langle v_L - v_S - v_R \rangle$ as a constraint. Therefore, lexical items as "*beautiful*", "*difficult*", "*pig*", and "*hate*", can be defined with such constraints as:
      · "beautiful", $ADJ_{[EH]} \Rightarrow TE\langle v_R \rangle \wedge \Rightarrow LSV\langle beauty \rangle$.
      · "difficult", $ADJ_{[EH]} \Rightarrow TE\langle v_R \rangle \wedge \Rightarrow LSV\langle complex \rangle$.
      · "pig", $NOUN_{[EH]} \Rightarrow TE\langle v_L \rangle \wedge \Rightarrow LSV\langle Pleasing - Likability \rangle$.
      · "hate", $VERB_{[EH]} \Rightarrow TE\langle v_L \rangle \wedge \Rightarrow LSV\langle Esteem \rangle$.
    * Moreover, it can be specified whether such lexical items are a prototypical or borderline evaluative expression, **always** with regards to the Linguistic

Semantic Variable (the semantic prime). If the evaluative head is prototypical with regards to the prime, it will be defined with the semantic constraint $[PROT_{(LSV_\alpha)}]$. If the evaluative head is borderline with regards to the prime, it will be defined with the semantic constraint of $[BORD_{(LSV_\gamma)}]$. For example:

- "beautiful", $ADJ_{[EH]} \Rightarrow TE\langle v_R\rangle \wedge \Rightarrow LSV\langle beauty\rangle [PROT_{(LSV_\alpha)}]$.
- "pig", $NOUN_{[EH]} \Rightarrow TE\langle v_R\rangle \wedge \Rightarrow LSV\langle cleanliness\rangle [BORD_{(LSV_\gamma)}]$.

- The last constraint is $X_{[EH]} \Rightarrow Sentiment_{TE-Judgment\langle v_L-v_S-v_R\rangle}$. Meaning that any category X as an evaluative head $X_{[EH]}$ requires a sentiment to be assigned $\Rightarrow$ Sentiment. Sentiment is characterized as a trichotomous expression of judgment $TE-Judgment\langle v_L-v_S-v_R\rangle$, such as $\langle negative - neutral - positive\rangle$. As with before, the sentiment can be prototypical or borderline, since it is considered that any evaluation can be potentially used in both ways [16]. For example:

  * "This song is _excellent_", it is interpreted as positive because the word "_excellent_" is prototypical positive: $ADJ_{[EH]} \Rightarrow Sentiment_{TE-Judgment\langle v_R\rangle}[PROT_{(S_\alpha)}]$
  * "This song is _excellent for deaf people_", it is used as a borderline negative because of the linguistic context: $X_{[EH]} \Rightarrow Sentiment_{TE-Judgment\langle v_L\rangle}[BORD_{(S_\gamma)}]$

Summarizing, in FNL we define an evaluative expression as a linguistic construction characterized by a lexical unit tagged as Evaluative Heading (EH), which is a subjective evaluation of an object. Evaluative heads have the following properties:

- They occur in each member of a triplet of expressions –an _evaluative trichotomy_–, such as _small-medium-big, short-medium-tall, beautiful-average-ugly_. A trichotomous expression is a characterization of a general semantic feature –a _semantic prime_– called a _linguistic semantic value_. Examples of these linguistic semantic values are _height, beauty, complexity, intelligence_, etc. Technically, we can refer to _features_ of a given object.
- They have a sentiment value expressed in terms of positive, negative degree. Sentiment is computed by ordering all the trichotomous expressions as protoypically negative v_L, protoypically medium v_S, protoypically positive v_R, and converting those to borderline sentiment when needed.
- FNL can characterize the interpretability of an evaluation even if it is subjective.

5. Towards a Lexicon of Evaluative Expressions

The formal characterization of evaluative expressions we have presented in this paper constitutes a robust mathematical model that is inapplicable without a lexicon of evaluative expressions. One of the main reasons for this is because linguistic categories, or part-of-speech, are not enough to identify an evaluative expression. It is, therefore, necessary to establish a procedure to find and classify evaluative expressions in trichotomous structures by identifying semantic primes and assigning values on a sentiment scale.

Taking into account the need to have a lexicon of evaluative expressions, we have carried out a first experiment, as a proof of concept, to identify trichotomic expressions in Spanish and English. We have performed a linguistic evaluation for FNL over a sentiment analysis corpus entirely build on evaluative words.

For our proof-of-concept we have chosen SO-CAL, lexicon. https://github.com/sfu-discourse-lab/SO-CAL. SO-CAL has two tagged sentiment corpus, one in Spanish, and another one in English. SO-CAL has three different lexicons for each language: (1) a lexicon for evaluative adjectives, (2) a lexicon for adverbs, and (3) a lexicon for intensifiers. SO-CAL lexicon was built through pairing of DSm tools. The corpus merely tagged each word with a value from -5 to $+5$ without further detail.

SO-CAL [46] was the best option for building up a lexicon with FNL, since it already has a classification of 11-point scale, which fits with the _theoretical extension_ of FNL (Figure 5).

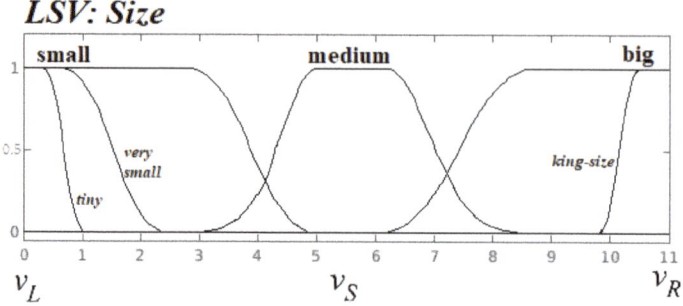

Figure 5. Trichotomous Expression of [*small-medium-big*] in FNL.

We have taken the SO-CAL lexicon and re-distributed its adjective-lexical units. We were interested in re-building the lexicon for characterizing evaluative adjectives with FNL providing more linguistic information for each EH, and re-distributing the sentiment polarities.

In order to built a lexicon underpinned in SO-CAL lexicon, two researchers elaborate *Linguistic Semantic Variables* (LSV) scales. One worked with an English Lexicon, and the other worked with a Spanish Lexicon. Without having contact with each other, they had to find the Prime Evaluative Heads (EHs) for each LSV scale. They have also to tag every adjective into an LSV scale, under a Prime EH.

The English adjective lexicon has 2827 lexical units, and the Spanish Adjective lexicon has 2049.

In Figure 6, the x-axis displays the sentiment polarity for each adjective cluster $[-5, 5]$. At the same time, the y-axis represents the number of words (in proportion from one corpus to the other) for each polarity set in the lexicon. Therefore, thanks to Figure 6, we confirm that SO-CAL does not demonstrate substantial differences in the classification of the adjectives' polarity in English and Spanish. Moreover, SO-CAL lexicon demonstrates a preference for medium values. Most of the adjectives are classified in the polarity of $[+2, +1]$, and $[-2, -1]$, leaving the 0 value just with one case, with the word *better*.

On the other hand, we re-classified the SO-CAL lexicon applying the criteria from FNL. We manually re-built half of the lexicon for both languages. We re-tagged 1419 adjectives for the English language, and 1549 adjectives for the Spanish Language. Therefore, it allowed us to evaluate FNL cross-linguistically.

After manually tagging each adjective under an LSV, and under a prime EH, the general result was that the evaluation values' distribution considerably changed in both FNL lexicons.

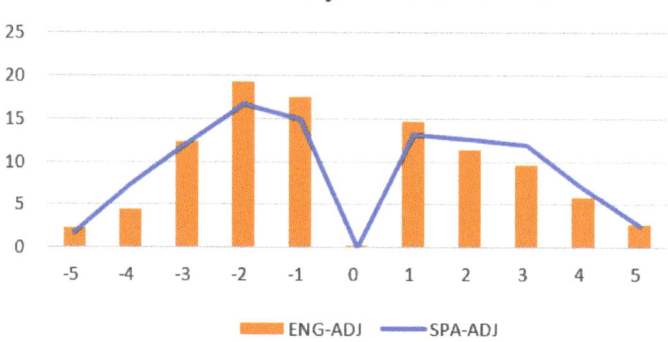

Figure 6. SO-CAL Adjectives by polarity.

In both Figures 7 and 8, the FNL lexicon preferred extreme values rather than middle ground values. However, both classifications coincide in leaving the very mid almost empty, with the exception that in Figure 7, it has a visible small peak in the mid value in contrast with Figures 6 and 8. These results can be explained by taking into consideration the following:

- Evaluation of the plots;
- Universality of the mid-value;
- Struggles with TE-adjective in LSV.

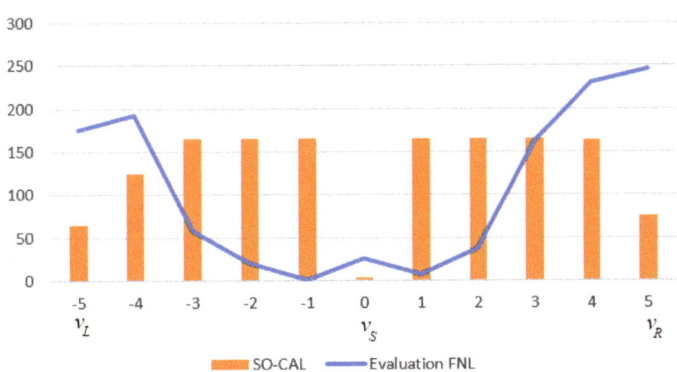

Figure 7. English Adjectives SO-CAL vs. FNL lexicon.

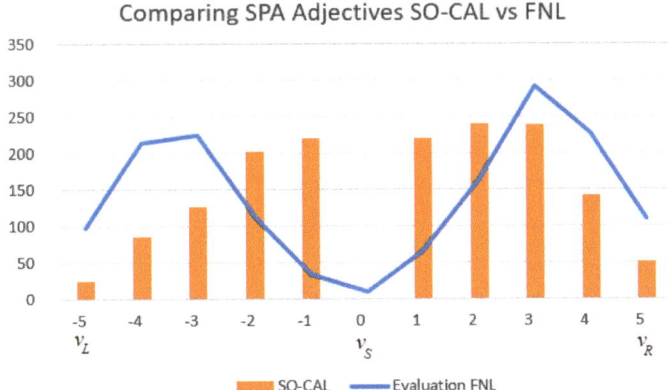

Figure 8. Spanish Adjectives SO-CAL vs. FNL lexicon.

Regarding *evaluating the plots*, it was more natural to classify the plots roughly either in v_L or v_R. Acknowledging the context was unnecessary to decide whether the word is placed in v_R or v_L or v_S. We conclude that the universality of FNL resides in the rough understanding of fuzzy prime EHs. Therefore, an EH is usually roughly either a v_R or v_L. However, when we try to fine-grain a particular value, we do need the linguistic context to clarify whether that word is a prototypical closer to the extremes of either v_L or v_R, or whether it is a lexical unit closer to the middle plot v_S, which intersections with v_L or v_R.

Regarding the *universality of the mid-value* v_S, it turns out that we could barely classify words in v_S. Most of them are general enough to be included in almost every LSV. Some examples are: *ordinary, average, average-sized, grey, central, medium, adult, similar, normal*. We jump into the following conclusion according to this phenomena. Mid words are the universal words in an evaluation scale. Because of their lack of markedness, they can flexibly be included in most of the scales as v_S, and they still keep the mid-meaning. Words in v_S can be considered the fuzziest semantics units since they do not show any clear-cut semantic traits, and yet they roughly fit in several LSV. Table 2, shows some TE-adjective with an *X* in the middle. *X* means that the mid-plot is still existing; however, there is no specific prime EH for such plot.

Table 2. Linguistic Semantic Variable (LSV) and Prime Evaluative Head (EH) tags in Fuzzy Natural Logic (FNL) Lexicon.

		FNL Tags in English and Spanish Lexicon			
LSV	Judgment	Esteem	Beauty	Size	Capability-Skills
Primes EH	⟨negative/bad-medium/normal-positive/good⟩	⟨hated-X-loved⟩	⟨ugly-X-beautiful⟩	⟨small/short-medium-big/long⟩	⟨disable-average-capable⟩
LSV	Complexity	Fear-Courage	Fullness	Indeterminacy	Intelligence
Primes EH	⟨simple-normal-complex⟩	⟨scared-X-brave⟩	⟨empty-X-full⟩	⟨blurred-X-clear⟩	⟨stupid-average-intelligent⟩
LSV	Generates-Interest	Pleasing-Likability	Proximity	Veridicality	Similarity-Usual
Primes EH	⟨boring-X-interesting⟩	⟨disgusting-X-pleasing⟩	⟨far-central-close⟩	⟨fake/false-X-real/truth⟩	⟨different-similar-usual⟩
LSV	Speed	Strength-Intensity	Temperature	Time-Lifetime	Worth-Value
Primes EH	⟨slow-medium-fast⟩	⟨weak/fragile-X-intense/strong⟩	⟨cold-X-hot⟩	⟨life/new/young/beggining- medium/adult-death/old/end⟩	⟨worthless-X-worthy⟩
LSV	Value (economical)				
Primes EH	⟨cheap-affordable-expensive⟩				

Finally, regarding *struggles with TE-adjective in LSV*, Table 2 displays the tags that have been found and agreed after manually tagging both the Spanish and English corpus.

We have found more satisfying those scales that classifies more specific LSV, such as *Fear-Courage, Esteem, Complexity, Capability-Skill, Temperature,* or *Value (economical)* in contrast with these scales with a more general LSV such as *Size, Beauty, Pleasing-Likability, Fullness*, and so on. Using hyper-generic LSV such as *Size*, or *Pleasing-Likability* can bias a fine-grained classification, since it is easy to fit too borderline cases into those scales. Examples of too borderline cases in the LSV of *Pleasing-Likability* would be: *happy, funny,* or *clean*. Assuming that all these words transmit a semantic value of something being *pleasing*. The LSV of *Fullness* is a particular one. This LSV is a mishmash of evaluative heads that cannot completely fit in other LSV without creating a specific scale with almost no adjectives. Some of the cases are: *informative* as (full of information), *illogic* as (empty of logic), words with the morpheme *full* like *help-ful* as (full of help), same with words with *-less* or *-free* like *painless* (empty of pain), and *dramafree* (empty of drama), among others.

Therefore, one of the significant challenges in the classification of an FNL lexicon is (1) not being too specific without reason during the extraction of LSVs, and (2) not pushing the fit of borderline cases in a hyper-generic LSV.

On the other hand, two other big struggles arise when tagging adjectives:

(1) It is complex to know under which LSV tag the adjective should be classified if we do not look at the linguistic context of the word,
(2) Some adjectives might be suspicious of being able to appear in more than one LSV scale. It is the case of *beautiful* which could be *Beauty*: *she is beautiful*, and/or *Pleasing-Likability*: *the food taste beautiful*, and/or *Judgment*: *it is a beautiful job*.

These two struggles brought us to implement a new feature in the FNL lexicon. We are implanting a sub-tag of prototypical LSV (LSV_α) and borderline LSV (LSV_γ). This tag is particular for each lexical unit. Which means that if a prime EH such as *beautiful* is a tag with LSV_α and LSV_γ does not necessarily imply that the lexical units under this prime will require the same specific tagging.

Table 3 displays how every particular lexical unit is tagged with prototypes, borderline, and sentiment features. In the case of Table 3, the TE-adjective membership coincides in the three cases. Let us go back to Table 2. We will realize how all the TE are prepared for automatically converging in most cases: v_L is usually equivalent for a prototypical negative, v_S is usually equivalent for a prototypical neutral, and v_R is usually equivalent for a prototypical positive.

Table 3. Example of a tagged lexical unit.

	Lexical Unit: *Beautiful*	
LSV_α	Beauty	v_R
LSV_γ	Judgment, Pleasing-Likability	v_R
Sentiment	positive	v_R

However, in some cases, some adjectives do need a dual tag for *prototype sentiment* (S_α) and *borderline sentiment* (S_γ). These are the example of words such as *sick*, *geek*, among others.

Table 4 provides two examples in which its lexical units have borderline cases in LSV and sentiment. Both cases clearly specify that the borderline case is activated only when a specific LSV is triggered. Therefore, within a particular context, *sick* would trigger S_γ in a sentence such as *Jordan's game is sick (skills)*. Same situation for *geek*. It would trigger a borderline sentiment in a sentence such as *I am a Netflix geek (generates-interest)*.

Table 4. Tagging lexical units with borderline cases.

	Lexical Unit: *Sick*			**Lexical Unit: *Geek***	
LSV_α	Judgment	v_L	LSV_α	Similarity-Usual	v_L
LSV_γ	Pleasing-Likability Capability-Skills	v_L v_R	LSV_γ	Intelligence	v_R
S_α	negative	v_L	S_α	negative	v_L
S_γ	positive {IFF capability-skills}	v_R	S_γ	positive {IFF generates-interest}	v_R

Therefore, thanks to FNL, it is only necessary to specify the borderline LSV_γ, and in which side of the fuzzy-graph the lexical unit will be in the borderline case (v_L or v_R).

We must mention that the borderline cases are challenging to extract. They arise as words with extreme variability, making it very difficult to know which exact word-pairing are borderline, and which borderline properties are triggering. The first step in our future work is working on an FNL lexicon with tagged-nouns. Therefore, we will be in a better position to evaluate the characterization of borderline cases.

Summing up, with this experimental work, we aimed:

- To evaluate the model's advantages and limitations when building a lexicon with FNL—improving the model with new linguistic properties.
- To check if it is feasible to convert adjectives from the semantics of the evaluative expression into an equivalent *positive-negative* value.
- To check in which cases the linguistic context is needed to classify an evaluative adjective in a LSV under a prime EH.

6. Conclusions

The model we have introduced takes into account the four main features when describing evaluative expressions:

- Firstly, evaluative linguistic expression's syntax, semantics, and sentiment are vague and gradient. The vagueness and gradience of its syntax and semantics are definable through the concepts of grammaticality and coercion.
- Secondly, the semantics of an evaluative linguistic expression can be associated with a semantic prime with a sentiment value. A real number cannot represent the extension of this prime most of the time. Therefore, it is necessary to consider an abstract extension representing an intensity similar to a Liker-t scale.
- Thirdly, every evaluative linguistic expression has a single lexical unit as an evaluative head. Such heads are not necessarily adjectives and have to be defined in terms of prototypical and borderline meanings.
- Lastly, syntactic structures that violate syntactic constraints detonate semantic intensions equivalent to a semantic prime. Therefore, variability does not compromise the final processing of the meaning of the construction.

Our findings and research provide the basis for creating a lexicon of evaluative expressions for future computational applications. Furthermore, we have expanded the trichotomous expressions, extracting new ones from a sentiment analysis corpus by experts.

Additionally, we outline the following aspects related to some of the theoretical criticism this model might receive.

- The linguistic mechanisms displayed in this paper to define evaluative expressions are mainly found in Sections 3 and 4. Those mechanisms are mainly the linguistic constraints from FPGr (Section 3) and selected concepts of Fuzzy Natural Logic (FNL) adapted to FPGr (Section 4), which consist of using the description of the trichotomous expression $\langle v_L, v_S, v_R \rangle$ as a constraint to define the semantics of an evaluative head. So, for example, some evaluative heads would have a constraint of $\langle VR \rangle$ that define its semantics, such as *"big"*, *"tall"*, *"complex"*, *"beautiful"*, and so on. In summary, the general approach to the model, its mechanisms such as the linguistic constraints, and its architecture are married to how FNL compute evaluative expressions.
- Evaluative expressions as a "component" are identified with a lexicon. Therefore, building a lexicon is fundamental for applying both FPGr and FNL to characterize evaluative expressions, mainly because an evaluative head does not have a fixed linguistic category.
- Our proposal is theoretical, including a proof-of-concept of how to build a lexicon of evaluative expressions. In the future, it is necessary to extend the tags on the evaluative expressions because it does need a lexicon to identify them, so their special fuzzy semantics can be characterized.
- Regarding whether the system prevents over-generation or not, FPGr and FNL system does not generate language, treebanks, or tree structures; they only characterize language. The generation of constraints or establishing the number of constraints for each language is up to each language's grammar or the experts configuring such grammar. If there is a problem with the overgeneration, it is not a fault of the system or the architecture. It is a fault of how the system was applied to define a particular language.
- The grammaticality of syntax is considered, as the gradient is a fundamental part of the FPGr. However, the notion of gradient grammaticality is not as "radical" as it might seem under the architecture of FPGr. First, as observed in Section 3, the grammar is defined as a set of constraints. Thus, the set of constraints represents the knowledge of grammar. These constraints can be prototypical; therefore, they are, for sure, constraints of a specific grammar, or borderline, so they are "sometimes" constraints of specific grammar. That is, they are marked, and they appear in specific contexts that are not prototypical. Therefore, the grammar is fuzzy because it considers both prototypical and borderline constraints and weighs them. Secondly, a natural

language input will have a degree of membership of a specific grammar according to the number of satisfied and violated constraints of such specific grammar. Therefore, the degree of grammaticality is defined as the degree of belongingness of an input, with regards to specific natural language grammar. For further information, see Torrens-Urrutia [19].

- FPGr and FNL do not consider extra-linguistic interpretations, since such a case would entail pragmatics. Therefore, it cannot generate from an input such as *"I love your food"* an output as: *"I love the fact that you cooked for me, regardless of the quality of the food"*. Such an interpretation is considered pragmatic due to extra-linguistic reasons, i.e., theory of politeness, context, intention, implicatures, etc. (see Figure 2). Semantics in FPGr only focuses on the essential part of the meaning, the most primitive one, that is why we defend the notion of semantic primes as a hypernym, and the system defines constraints of evaluations regarding a prime. These constraints are $\langle v_R, v_S, v_L \rangle$.

7. Future Work

The future work for Fuzzy Natural Logic involves expanding the *fundamental trichotomous expressions and LSVs*. For such a task, it will be necessary to either explore computational techniques which can automatically extract the LSV, or to perform a systematisation of surveys in which a significant number of native speakers with linguistic knowledge of their language perform classifications of words. Regardless of the method, the final objective should be the creation of an ontology of the evaluative meaning based on the feature descriptions that this work pointed out.

Figure 9 shows an ontology made *ad hoc* as a guidance for a future design of a computational one. On the top, at the same level, we find both the *Prime Sentiment*, and *Prime semantics*, which will typically share polarities; that is, v_L will be prototypically negative. At the same time, v_R will be prototypically positive. From the most abstract notion of prime, we will find the next level, at which we find the *fundamental trichotomous expressions* with their *LSVs*. The closer we get to the bottom, the more metaphorical evaluations will be found. The prototypical relations are marked with a straight line, while fractured lines denote the borderline relations. The lines will state the semantic distribution between the prime and the related words if they have a relationship to v_L, v_S, or v_R. For example, *Einstein* has a borderline relationship of being a metaphorical evaluation with a semantic polarity of v_R in the LSV prime of *intelligence*, which means that it will be included in the fuzzy set of the $TE - EH$ of *intelligent*. The ontology must be able to define evaluative words as a combination of several *fundamental trichotomous expressions*, such as the case of *cute*. Typically, something *cute* is referred for something closer to be *pretty*, but at the same time, it shares traits with something being *quite small*. We would not say that a *big truck* is *cute*, but a *cartooney small toy truck* could be *cute*. Independently of how right is this semantic analysis, the main advantage of such an ontology is that it provides 'explainability', and the particular cases can be changed and improved. Thus, if a semantic relation is inaccurate, it could be fixed. Some words might have a lot of semantic relationships. It is a case of *do-it-yourself*. This lexical unit operates as an evaluation, providing a lot of different borderline meanings.

Creating several ontologies would help linguistics evaluate complexity of the syntax and semantics across languages for future experiments and linguistic applications in explainable AI. That is, computing with constraints how similar the meanings between languages are, elaborate ontologies of the history of language through the evolution of importance, and the evaluation of the linguistic universality in the semantic domain, among others.

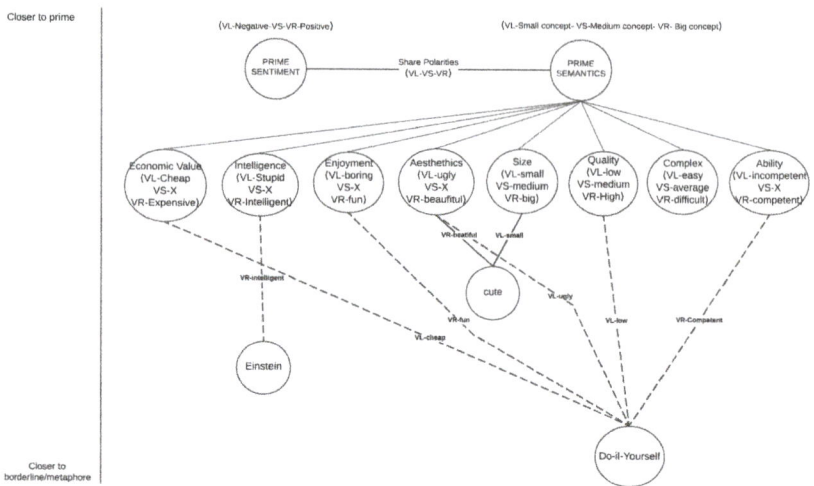

Figure 9. Ontology of evaluative expressions.

Author Contributions: All authors have contributed equally. Conceptualization, A.T.-U., V.N. and M.D.J.-L.; Formal analysis, A.T.-U., V.N. and M.D.J.-L.; Writing—original draft, A.T.-U., V.N. and M.D.J.-L.; Writing—review & editing, A.T.-U., V.N. and M.D.J.-L. All authors have read and agreed to the published version of the manuscript.

Funding: This paper has been supported by the project CZ.02.2.69/0.0/0.0/18_053/0017856 "Strengthening scientific capacities OU II".

Data Availability Statement: https://github.com/sfu-discourse-lab/SO-CAL (accessed on 10 January 2022).

Acknowledgments: We also want to thank Maite Taboada for her collaboration and support during this research and Eduard Mir Neira for his work with the Spanish Lexicon LSV scales.

Conflicts of Interest: The authors declare no conflict of interest.

References

1. Novák, V. The Concept of Linguistic Variable Revisited. In *Recent Developments in Fuzzy Logic and Fuzzy Sets*; Sugeno, M., Kacprzyk, J., Shabazova, S., Eds.; Springer: Berlin/Heidelberg, Germany, 2020; pp. 105–118.
2. Novák, V. Fuzzy Logic in Natural Language Processing. In Proceedings of the International Conference FUZZ-IEEE, Naples, Italy, 9–12 July 2017.
3. Novák, V.; Perfilieva, I.; Dvořák, A. *Insight Into Fuzzy Modeling*; Wiley & Sons: Hoboken, NJ, USA, 2016.
4. Novák, V. What is Fuzzy Natural Logic. In *Integrated Uncertainty in Knowledge Modelling and Decision Making*; Huynh, V., Inuiguchi, M., Denoeux, T., Eds.; Springer: Berlin/Heidelberg, Germany, 2015; pp. 15–18.
5. Novák, V. Fuzzy Natural Logic: Towards Mathematical Logic of Human Reasoning. In *Fuzzy Logic: Towards the Future*; Seising, R., Trillas, E., Kacprzyk, J., Eds.; Springer: Berlin/Heidelberg, Germany, 2015; pp. 137–165.
6. Novák, V. Evaluative linguistic expressions vs. fuzzy categories? *Fuzzy Sets Syst.* **2015**, *281*, 81–87. [CrossRef]
7. Novák, V. Mathematical Fuzzy Logic: From Vagueness to Commonsese Reasoning. In *Retorische Wissenschaft: Rede und Argumentation in Theorie und Praxis*; Kreuzbauer, G., Gratzl, N., Hielb, E., Eds.; LIT-Verlag: Wien, Austria, 2008; pp. 191–223.
8. Lakoff, G. Linguistics and natural logic. *Synthese* **1970**, *22*, 151–271. [CrossRef]
9. Novák, V.; Murinová, P.; Boffa, S. On the properties of intermediate quantifiers and the quantifier "MORE-THAN". In *Information Processing and Management of Uncertainty in Knowledge-Based Systems, Part III*; Medina, J., Ojeda-Aciego, M., Verdegay, J.L., Pelta, D.A., Cabrera, I.P., Bouchon-Meunier, B., Yager, R.R., Eds.; Springer Nature: Zürich, Switzerland, 2020; pp. 159–172.
10. Murinová, P.; Novák, V. The theory of intermediate quantifiers in fuzzy natural logic revisited and the model of "Many". *Fuzzy Sets Syst.* **2020**, *388*, 56–89. [CrossRef]
11. Novák, V.; Murinová, P. A formal model of the intermediate quantifiers "A few, Several, A little". In *Fuzzy Techniques: Theory and Applications*; Kearfott, R., Batyrshin, I., Reformat, M., Ceberio, M., Kreinovich, V., Eds.; Springer: Cham, Switzerland, 2019; pp. 429–441.
12. Nguyen, L.; Novák, V. Forecasting seasonal time series based on fuzzy techniques. *Fuzzy Sets Syst.* **2019**, *361*, 114–129. [CrossRef]

13. Novák, V.; Pavliska, V.; Perfilieva, I.; Stepnicka, M. F-transform and Fuzzy Natural logic in Time Series Analysis. In Proceedings of the EUSFLAT Conference, Milano, Italy, 11–13 September 2013.
14. Novák, V. Linguistic characterization of time series. *Fuzzy Sets Syst.* **2016**, *285*, 52–72. [CrossRef]
15. Novák, V.; Perfilieva, I.; Dvořák, A.; Chen, G.; Wei, Q.; Yan, P. Mining pure linguistic associations from numerical data. *Int. J. Approx. Reason.* **2008**, *48*, 4–22. [CrossRef]
16. Torrens Urrutia, A.; Jiménez-López, M.D.; Novák, V. Fuzzy Natural Logic for Sentiment Analysis: A Proposal. In Proceedings of the International Symposium on Distributed Computing and Artificial Intelligence, Salamanca, Spain, 6–8 October 2021; Springer: Berlin/Heidelberg, Germany, 2021; pp. 64–73.
17. Torrens Urrutia, A. Towards a fuzzy grammar for natural language grammars. In *Recerca en Humanitats 2018*; Publicacions URV: Reus, Spain, 2018; pp. 271–282.
18. Torrens Urrutia, A. An Approach to Measuring Complexity with a Fuzzy Grammar & Degrees of Grammaticality. In Proceedings of the Workshop on Linguistic Complexity and Natural Language Processing, Santa Fe, NM, USA, 25 August 2018; pp. 59–67.
19. Torrens Urrutia, A. A Formal Characterization of Fuzzy Degrees of Grammaticality for Natural Language. Ph.D. Thesis, Universitat Rovira i Virgili, Tarragona, Spain, 2019.
20. Torrens Urrutia, A.; Jiménez-López, M.D.; Brosa-Rodríguez, A. A Fuzzy Approach to Language Universals for NLP. In Proceedings of the IEEE International Conference on Fuzzy Systems (FUZZ-IEEE), Virtual, 11–14 July 2021; pp. 1–6.
21. Hemmatian, F.; Sohrabi, M.K. A survey on classification techniques for opinion mining and sentiment analysis. *Artif. Intell. Rev.* **2019**, *52*, 1495–1545. [CrossRef]
22. Yadav, A.; Vishwakarma, D.K. Sentiment analysis using deep learning architectures: A review. *Artif. Intell. Rev.* **2020**, *53*, 4335–4385. [CrossRef]
23. Taboada, M. Sentiment Analysis: An Overview from Linguistics. *Annu. Rev. Linguist.* **2016**, *2*, 325–347. [CrossRef]
24. Baccianella, S.; Esuli, A.; Sebastiani, F. Sentiwordnet 3.0: An enhanced lexical resource for sentiment analysis and opinion mining. In Proceedings of the LREC, Valletta, Malta, 17–23 May 2010; Volume 10, pp. 2200–2204.
25. Mohammad, S.; Dunne, C.; Dorr, B. Generating high-coverage semantic orientation lexicons from overtly marked words and a thesaurus. In Proceedings of the 2009 Conference on Empirical Methods in Natural Language Processing, Singapore, 6–7 August 2009; pp. 599–608.
26. Wilson, T.; Wiebe, J.; Hoffmann, P. Recognizing contextual polarity in phrase-level sentiment analysis. In Proceedings of the Human Language Technology Conference and Conference on Empirical Methods in Natural Language Processing, Vancouver, BC, Canada, 6–8 October 2005; pp. 347–354.
27. Wiebe, J.; Wilson, T.; Bruce, R.; Bell, M.; Martin, M. Learning subjective language. *Comput. Linguist.* **2004**, *30*, 277–308. [CrossRef]
28. Blache, P. Representing syntax by means of properties: A formal framework for descriptive approaches. *J. Lang. Model.* **2016**, *4*, 183–224. [CrossRef]
29. Universal Dependency Corpora. Available online: https://universaldependencies.org/ (accessed on 1 September 2021).
30. Blache, P.; Rauzy, S.; Montcheuil, G. MarsaGram: An excursion in the forests of parsing trees. In Proceedings of the 12th Language Resources and Evaluation Conference, LREC 2020, Marseille, France, 11–16 May 2020; Volume 10, p. 7.
31. Manca, V.; Jiménez-López, M.D. GNS: Abstract Syntax for Natural Languages. *Triangle* **2012**, *8*, 55–79. [CrossRef]
32. Fillmore, C.J. The mechanisms of "construction grammar". In Proceedings of the Annual Meeting of the Berkeley Linguistics Society, 13–15 February 1988; Volume 14, pp. 35–55. Available online: http://journals.linguisticsociety.org/proceedings/index.php/BLS/article/viewFile/1794/1566 (accessed on 10 January 2022).
33. Goldberg, A.E. *Constructions at Work: The Nature of Generalization in Language*; Oxford University Press on Demand: Oxford, UK, 2006.
34. Fillmore, C.J.; Baker, C. A frames approach to semantic analysis. In *The Oxford Handbook of Linguistic Analysis*; Oxford University Press: Oxford, UK, 2010.
35. Goldberg, A. Verbs, constructions and semantic frames. In *Syntax, Lexical Semantics, and Event Structure*; Oxford University Press: Oxford, UK, 2010; pp. 39–58.
36. Müller, G. Optimality, markedness, and word order in German. *Linguistics* **1999**, *37*, 777–818. [CrossRef]
37. Wierzbicka, A. *Semantics: Primes and Universals: Primes and Universals*; Oxford University Press: Oxford, UK, 1996.
38. Erk, K.; Padó, S.; Padó, U. A flexible, corpus-driven model of regular and inverse selectional preferences. *Comput. Linguist.* **2010**, *36*, 723–763. [CrossRef]
39. Baroni, M.; Lenci, A. Distributional memory: A general framework for corpus-based semantics. *Comput. Linguist.* **2010**, *36*, 673–721. [CrossRef]
40. Greenberg, C.; Sayeed, A.; Demberg, V. Improving unsupervised vector-space thematic fit evaluation via role-filler prototype clustering. In Proceedings of the 2015 Conference of the North American Chapter of the Association for Computational Linguistics Human Language Technologies, Denver, CO, USA, 31 May–5 June 2015; ACL: Denver, CO, USA, 2015; pp. 21–31.
41. Santus, E.; Chersoni, E.; Lenci, A.; Blache, P. Measuring thematic fit with distributional feature overlap. *arXiv*, **2017**, arXiv:1707.05967.
42. Crystal, D. On keeping one's hedges in order. *Engl. Today* **1988**, *4*, 46–47. [CrossRef]
43. Layman, L. Reticence in oral history interviews. In *The Oral History Reader*; Routledge: London, UK, 2015; pp. 234–252.
44. Islam, J.; Xiao, L.; Mercer, R.E. A lexicon-based approach for detecting hedges in informal text. In Proceedings of the 12th Language Resources and Evaluation Conference, Marseille, France, 11–16 May 202; pp. 3109–3113.

45. Hassan, A.S.; Said, N.K.M. A pragmatic study of hedges in American political editorials. *Int. J. Res. Soc. Sci. Humanit.* **2020**, *10*. [CrossRef]
46. Taboada, M.; Brooke, J.; Tofiloski, M.; Voll, K.; Stede, M. Lexicon-based methods for sentiment analysis. *Comput. Linguist.* **2011**, *37*, 267–307. [CrossRef]

Article
Preimage Problem Inspired by the F-Transform

Jiří Janeček [1,*,†] and Irina Perfilieva [2,*,†]

1. Department of Mathematics, Faculty of Science, University of Ostrava, 30. dubna 22, 701 03 Ostrava, Czech Republic
2. Institute for Research and Applications of Fuzzy Modeling, University of Ostrava, 30. dubna 22, 701 03 Ostrava, Czech Republic
* Correspondence: jiri.janecek@osu.cz (J.J.); irina.perfilieva@osu.cz (I.P.)
† These authors contributed equally to this work.

Abstract: In this article, we focus on discrete data processing. We propose to use the concept of closeness, which is less restrictive than a metric, to describe a certain relationship between objects. We establish a fuzzy partition of a given set of objects in a way that admits a closeness space to emerge. The fuzzy (F-) transform is a tool that maps objects with common characteristics to the same discrete image—the direct F-transform. We are interested in the inverse (preimage) problem: How can we describe the class of all functions mapped onto the same direct F-transform? In this manuscript, we focus on this preimage problem, formulated accordingly. Its solution is presented from three different points of view and shows which functions belong to the same class determined by a given image (by the direct F-transform). Conditions under which a solution to the preimage problem is given by the inverse F-transform over the same fuzzy partition, or by transforming a given image using a new system of basic functions, are formulated. The developed theory contributes to a better understanding of ill-posed problems that are typical for machine learning. The appendix contains illustrative numerical examples.

Keywords: closeness; closeness matrix; closeness space; function similarity; fuzzy partition; fuzzy transform; preimage problem; singular value decomposition

MSC: 15A29

1. Introduction

In this paper, we are focused on spaces with closeness that are characterized using weighted graphs, as in [1], or local neighborhoods of elements as in [2]. Formally, closeness can be described by the graph adjacency matrix that contains weights—a higher value of weight is assigned to the edge that connects closer objects. Those matrices were always square as they contained mutual closeness values of all objects in the dataset.

In our previous research [1,3], the notion of closeness was used in the new approach to dimensionality reduction for which fuzzy transform (F-transform for short, [4]) proved to be useful. Otherwise, the concept of closeness can be observed in multiple contexts and under different names, where it serves auxiliary purposes. For example, in [5], the phase space of a dynamical system is described by a network (graph) where each point represents one state of the system and closeness-describing weights (between two points) are equal to the frequency with which a transition occurred between the two states. There exists a related concept of closeness centrality (aggregation of closeness-describing weights with respect to all neighbors except oneself computed by arithmetic or harmonic mean) that describes the data density around one particular point in a network, used, e.g., in [6]. In the area of image processing (image compression, image segmentation, image retrieval, e.g., [7], etc.), the related concept of proximity space is widely used. It originated from [8] and then evolved—currently, it is used, e.g., to describe the similarity between pixels or patches.

In this research, we consider closeness determined by a fuzzy partition of a universe of discourse. This particular space structure is used in the theory of F-transforms [4,9]. The notion of F-transform was introduced in [4], where we explained modelling with fuzzy IF-THEN rules as a specific transformation. From this point of view, F-transform bridges fuzzy modelling and the theory of linear (in particular, integral) transforms. The generalization to a higher degree version was proposed in [9]. Generally speaking, F-transform performs a transformation of the original universe of functions into a universe of their skeleton models (vectors or matrices of F-transform components), for which further computations are easier. In [4], the approximation property of F-transform was described, and in [10], the effect of the shapes of basic functions on the approximation quality was demonstrated. F-transform has many other useful properties and great potential in various applications, such as special numerical methods as well as solutions to ordinary and partial differential equations with fuzzy initial conditions, mining dependencies from numerical data, signal processing, compression and decompression of images [11] and image fusion.

Among the recent applications of F-transform, we refer to an image classification problem, in which data is scarce [12], a numerical solution to fuzzy integral equations [13] and improving the JPEG compression algorithm in the cases where a high ratio compression is required [11].

Similarly to F-transform, which, in its direct phase, aggregates a large number of functional values of all points into a small number of components associated with selected nodes, we consider that within our (possibly large) dataset, there are several selected points (nodes) with known closeness values with respect to all the other points and that closeness among other points is undefined, which gives rise to a rectangular closeness matrix that describes the closeness space. Nodes can be thought of as the most prominent data points because the dataset can be sparsely represented as a collection of their neighborhoods.

Following the methodology of F-transform, we aim to demonstrate that a partial knowledge about the mutual position of all points in the dataset is sufficient for a successful retrieval of a discrete function (up to non-significant differences that distinguish particular functions from each other within the same equivalence class) from its F-transform components. Thus, we present a not just theoretical tool for demarcating the set of such functions that can be mutually replaced without losing an important feature in the space.

This allows us to express similarity between functions defined on spaces with closeness. In specific cases, a representative function is given by the inverse F-transform. Moreover, we contribute to the general theory of F-transforms in aspects described in this section. The task of reconstructing an element of an input space (or finding an approximate solution if it does not exist) based on (a lower-dimensional representation of) an element of a feature space has been studied in various machine learning domains, e.g., in kernel-based methods (e.g., [14] where the nonlinear dimensionality reduction is performed using the kernel trick on an input image and the task is to recover the image from its denoised version in the feature space, or [15] where algebraic tools based on the simple relationship of distances in the input space and in the feature space induced by a kernel are used to find the preimage of a feature vector) and graph-based methods (where the task is to recover structured data from a single point of a feature space, e.g., to find a representative graph based on features of a found data cluster). In these applications, the concept of preimage problem was already used. This paper is focused on finding all functions that share a set of features (given by the F-transform components) utilizing the above-described concept. We do not consider that these features might be noisy. On the other hand, in [16], we used a similar initial setting of the input space and we assumed that the input signal (function placed in this space) is noisy in a certain sense. The signal was processed in the same manner (based on closeness) and we proposed how to denoise the signal by finding the appropriate closeness parameters. The inverse F-transform is known to reduce the noise of the input signal (provided a suitable fuzzy partition) as was in the continuous case shown, e.g., in [17].

As explained in [18], preimage problems are useful not only in (pattern) denoising and Kernel Dependency Estimation but also in signal compression, where the kernel technique serves as an encoder and the preimage technique as a decoder. Moreover, in [18], a technique to learn the preimage without the need to solve a difficult nonlinear optimization problem is presented. Additionally, in [15], the authors consider that a noisy pattern is mapped to a feature space using the Kernel Principal Component Analysis and then the approximate preimage is found using their technique. To summarize the main stream of preimage problems in machine learning, we can say that a nonlinear optimization, nonlinear iteration, or learning method is used to find a function (the space of all possible functions comprises an input space), s.t. its image in a specified direct mapping has the minimal squared distance from the given point in a feature space. The existence of a precise solution is considered to be a coincidence. The purposes of computing the preimage in machine learning are various (to denoise a pattern, reconstruct a signal, compress an image, find a representative example of a data cluster, or learn a general mapping between the input space and the feature space) but all of them are highly application-oriented, and, hence, the initial settings are adjusted accordingly. This limits their transferability to another task.

In contrast, our approach pays close attention to the structure of the input space and this structure is quite flexible. We are focused on finding the precise solution (we ensure that it always exists) to the special case of the preimage problem. Using the assumption of the finiteness of the universe and endowing the input space with a fuzzy partition and the corresponding closeness, we solve it by the means of linear algebra. The consequent mapping forms a special case of direct mappings mentioned above. Since our theoretical work is not aimed at solving any specific task, we compute the whole class (usually infinite) from which we do not need to choose a particular solution (all of them are equivalent).

In Section 2, we give the details about the notion of closeness considered in this article. After that, the fuzzy partition and transform are recalled and modified for the discrete case. We show that a certain fuzzy partitioned space (space with a fuzzy partition) is a special case of the space with closeness. Section 3 discusses the formulation of the main topic of the paper—called preimage problem—that takes place in the space introduced in Section 2. Preimage problem belongs to basic problems in algebra (e.g., in [19]) where it is considered between certain two structured spaces (we propose to use the space with closeness and the space of scaled F-transform components). The solution to this problem is described in three different ways. Theorem 1 is based on a commonly used solution to the set of algebraic equations. In Section 4, we examine the conditions when the inverse F-transform forms the solution to this problem. In Section 5, we propose a technique consisting of checking whether a solution to the preimage problem can be obtained by a certain transformation of the given element of the space of scaled F-transform components given by a new set of fuzzy partition units. Appendix A provides the numerical examples related to four preceding sections, except for the Conclusions (Section 6).

Relationship of Closeness, Metric and Similarity

A space with closeness (X, w) is more general than a metric space (X, ρ) as they share the axioms of symmetry and non-negativity but metric is more restrictive—it requires two additional axioms ($\rho(x, y) = 0$ iff $x = y$ and the triangle inequality $\rho(x, y) \leq \rho(x, z) + \rho(y, z)$ for all $x, y, z \in X$). As closeness is a relaxed version of a reciprocal distance metric, it is more suitable for the description of data with graph structure (where the triangle inequality is generally non-enforceable). Another example of a context where closeness is more suitable is formed by the data that is assumed to lie on a topological manifold as there is no straightforward way how to establish a metric there. Similarly to metric, closeness encodes mutual relations between data points.

There is a close connection between closeness and similarity measures—each of these concepts applies in different contexts. An algebraic, fuzzy, or probabilistic background is usually assumed as a similarity requirement. There is no standard axiomatization of its definition. Similarly to closeness, a similarity value is higher for more similar objects (that are, e.g., more correlated, have a higher intersection, smaller cross-entropy, or belong to the same cluster), its values can be negative, it can be non-symmetric, etc. Similarity spaces emerging in various applied fields are even more general than spaces with closeness. A solution to a particular type of problem is often associated with a certain type of similarity measure.

This explains why we introduce the notion of closeness and why we give it a preference over that of metric (too-restrictive axioms) and similarity (too-loose axiomatization). This good trade-off allows us to express various established concepts, e.g., of graph theory (edge weights used, e.g., in the dimensionality reduction technique of Laplacian eigenmaps [2]), fuzzy logic (values of biresiduum of truth values of two formulae), clustering problems (class membership degrees in, e.g., KNN clustering algorithm), etc., in terms of closeness and incorporate a theoretical apparatus upon that. Therefore, as the closeness values can be extracted from the data in various ways, the closeness space creates a platform that enables both data-driven and model-driven approaches to be utilized.

2. Preliminaries

In this section, we describe the mathematical background by giving some basic definitions and properties. Throughout the manuscript, all settings are discrete.

2.1. Closeness Space

We are interested in the notion of closeness on finite sets that, as described below, agrees with [1,2]. To emphasize its general applicability, we express it using the language of classical set theory. We show that it can be also expressed in the languages of fuzzy set theory and graph theory.

Definition 1 (Closeness). *Let $X = \{x_i \mid i = 1, \ldots, L\}$ be a set, $L \in \mathbb{N}$, then* closeness *on X is any symmetric, non-negative function $w \colon X^2 \to \mathbb{R}$, where symmetry means $\forall x, y \in X \colon w(x,y) = w(y,x)$ and $X^2 = X \times X = \{(x,y) \mid x, y \in X\}$ is the Cartesian product of X with itself. The pair (X, w) is called a* space with closeness *or simply a* closeness space.

We provide the following semantic interpretation of closeness: objects that are closer in a certain sense (as explained in the Introduction, in some cases, metric cannot be defined) have a greater closeness value (with respect to natural order on $\mathbb{R}$) than the less-close ones. Non-close objects have a closeness value equal to zero, and, vice versa, if the closeness value of a pair of objects is greater, we consider these objects closer in this sense than a pair of objects with a smaller closeness value. The closeness value hence quantifies a certain quality of mutual alikeness between objects.

Example 1. *Let $X = \{0, 10, 20, 30, 40, 50\}$ and for all $x_i, x_j \in X \colon w(x_i, x_j) = \frac{1}{1+|x_i-x_j|}$, then w is closeness on X and (X, w) is a space with closeness.*

Example 2. *Let ρ be a metric and (X', ρ) a metric space. Let moreover, $X \subseteq X'$ and for all $x_i, x_j \in X$, $w_1(x_i, x_j) = \frac{1}{1+\rho(x_i, x_j)}$ and $w_2(x_i, x_j) = e^{-\rho(x_i, x_j)}$ hold true. Then, (X, w_1) and (X, w_2) are closeness spaces.*

Definition 2 ((Weakly) Reflexive Closeness). *Let $X = \{x_i \mid i = 1, \ldots, L\}$ be a set, $L \in \mathbb{N}$, and $w \colon X^2 \to \mathbb{R}$ be closeness on X. Then, w is* weakly reflexive closeness *on X if there exists a positive real number $\overline{w}$, s.t. $\forall x, y \in X \colon w(x,y) \leq w(x,x) = \overline{w}$. If, moreover, $\overline{w} = 1$, w is called* reflexive closeness *on X. The pair (X, w) is then a* space with (weakly) reflexive closeness *or simply a* (weakly) reflexive closeness space.

Remark 1. *In this paper, we consider only finite closeness domain X, hence there always exists a constant $m = \max\{w(x_i, x_j) \mid (x_i, x_j) \in X^2\}$, in case $1 < m < +\infty$, it would be more appropriate to define strong reflexivity by $\overline{w} = m$ but later, we work with closeness $w \colon X^2 \to [0,1]$.*

Example 2 contains two reflexive closeness-defining functions. Examples of weakly reflexive closenesses derived from an arbitrary metric ρ include $\frac{1}{2+\rho(x_i,x_j)}$ and $\frac{1}{2}e^{-\rho(x_i,x_j)}$, both with $\overline{w} = \frac{1}{2}$.

Let (X, w) be a space with closeness where $X = \{x_i \mid i = 1, \ldots, L\}$ is a set, $L \in \mathbb{N}$. Let us denote

$$\forall x_i, x_j \in X: w_{ij} = w(x_i, x_j), \qquad (1)$$

and consider a matrix that contains the closeness values

$$W = (w_{ij})_{i,j=1}^{L} \in \mathbb{R}^{L \times L}, \qquad (2)$$

then the matrix W is called the *matrix of closeness on X* or simply the *closeness matrix*. Functional values of closeness w for the elements of X^2 can be arranged into a matrix W—the matrix of closeness on X.

2.2. Fuzzy Set, Relation and Partition

Definition 3 (Fuzzy Set). *Let X be a non-empty* universe *and $K: X \to [0, 1]$ be a function. Fuzzy set K is the set of pairs $\{(x, K(x)) \mid x \in X\}$. The function K is a* membership function *and its functional value $K(x)$ is called the* membership degree *of the element x to K. Fuzzy set is conventionally identified with its membership function. We call it* fuzzy subset *of X which is denoted by $K \subsetneq X$. We say that the point $x \in X$ is* covered *by the fuzzy set K if $K(x) > 0$.*

Remark 2. *Note that if $K: X \to \{0, 1\}$, then K is a characteristic function of an ordinary set. The membership function $K: X \to [0, 1]$ is a generalization of the characteristic function of the set $K \subseteq X$ given by a function $\chi_K: X \to \{0, 1\}$. If the universe of a (fuzzy) set is clear, we can talk about (fuzzy) set K without referring to it.*

Definition 4 (Fuzzy Relation). *Let V and X be non-empty sets, then* binary fuzzy relation *r is any fuzzy subset of $V \times X$, i.e., $r \subsetneq V \times X$.*

Hence, the binary fuzzy relation r is identified with a membership function $r: V \times X \to [0, 1]$.

Definition 5 (Convex Fuzzy Set). *Let $X \subset \mathbb{R}$ and $K \subsetneq X$. Then K is a* convex fuzzy set*, if $\forall x_c, x_d, x_e \in X, x_c < x_d < x_e: K(x_d) \geq \min\{K(x_c), K(x_e)\}$.*

Following [4], we recall the notion of fuzzy partition and modify it for a discrete case. Consecutively, we set fuzzy transform in Section 2.6.

Definition 6 (Fuzzy Partition of a Real Interval with Nodes). *Let $[y_0, y_{l+1}]$ be a real interval and $y_1, \ldots, y_l \in \mathbb{R}$, s.t. $y_0 < y_1 < y_2 < \ldots < y_l < y_{l+1}$ and $0 < l < +\infty$. Then the* fuzzy partition *of $[y_0, y_{l+1}]$ is given by a set of fuzzy sets $\{A_1, \ldots, A_l\}$, if:*

1. *$\forall i = 1, \ldots, l: \quad A_i: [y_0, y_{l+1}] \to [0, 1]$,*
2. *for all $i = 1, \ldots, l: A_i$ is continuous with respect to topologies on $[y_0, y_{l+1}], [0, 1] \subset \mathbb{R}$ induced by the natural topology on $\mathbb{R}$,*
3. *$\forall i = 1, \ldots, l: \quad A_i(x) > 0$ iff $x \in (y_{i-1}, y_{i+1})$,*
4. *for all $i = 1, \ldots, l: A_i$ strictly increases in $(y_{i-1}, y_i]$,*
5. *for all $i = 1, \ldots, l: A_i$ strictly decreases in $[y_i, y_{i+1})$.*

Elements of the set $\{A_1, \ldots, A_l\}$ are called basic functions *(fuzzy partition units). Based on condition 3., $A_1, \ldots, A_l$ are associated with points $y_1, \ldots, y_l$, respectively. These points are called* nodes *of the fuzzy partition.*

In Definition 6, conditions 3–5 ensure that each basic function is a convex fuzzy set. Moreover, condition 3 ensures that each basic function covers at least one point. This might not hold in the case of a discrete universe which motivated the following definition.

Definition 7 (Sufficient Density with respect to Fuzzy Partition). *Let X be a set, s.t. $X \subset [y_0, y_{l+1}] \subset \mathbb{R}$, and $0 < l < +\infty$. Let $\mathcal{P} = \{A_1, \ldots, A_l\}$ be a fuzzy partition of the interval $[y_0, y_{l+1}]$. Then X is* sufficiently dense *with respect to $\mathcal{P}$, if $\forall i = 1, \ldots, l \, \exists x \in X \colon A_i(x) > 0$.*

Definition 7 states that a set is sufficiently dense with respect to a given fuzzy partition if every basic function covers at least one point of that set. This notion allows us to define the fuzzy partition of a discrete set.

Definition 8 (Fuzzy Partition of a Discrete Subset of a Real Interval with Nodes). *Let $X = \{x_1, \ldots, x_L\}$ be a set, s.t. $X \subset [y_0, y_{l+1}] \subset \mathbb{R}$, and $0 < l \leq L < +\infty$. Let $\mathcal{P}' = \{A'_1, \ldots, A'_l\}$ be a fuzzy partition of the interval $[y_0, y_{l+1}]$ with nodes in $Y = \{y_1, \ldots, y_l\} \subseteq X$ and X be sufficiently dense with respect to $\mathcal{P}'$. Then $\mathcal{P} = \{A_1, \ldots, A_l\}$ is the* fuzzy partition of X with nodes in Y, *if $\forall i = 1, \ldots, l \colon A_i = A'_i|_X$, i.e., if each basic function is replaced by its restriction on X.*

Remark 3. *Definition 8 can be extended in the converse direction: let $X = \{x_1, \ldots, x_L\}$ be a set, s.t. $X \subset [y_0, y_{l+1}] \subset \mathbb{R}$, $Y = \{y_1, \ldots, y_l\} \subseteq X$ and $0 < l \leq L < +\infty$. Let $\mathcal{P} = \{A_1, \ldots, A_l\}$, s.t. $\forall i = 1, \ldots, l \colon A_i \subsetneq X \, \& \, A_i(y_i) > 0$. Then $\mathcal{P}$ will be also called the fuzzy partition of X with nodes in Y, if there exists a fuzzy partition $\mathcal{P}' = \{A'_1, \ldots, A'_l\}$ with nodes in Y, s.t. X is sufficiently dense with respect to $\mathcal{P}'$ and for each $i = 1, \ldots, l$, A'_i is a continuous extension of A_i on $[y_0, y_{l+1}]$, i.e., $A'_i \subsetneq [y_0, y_{l+1}]$, A'_i is continuous and $A_i = A'_i|_X$.*

Below, we show how the space with closeness can be represented by a fuzzy partitioned space and by a weighted graph-structured space that, satisfying certain conditions, are examples of closeness spaces.

2.3. Closeness as Fuzzy Relation and Closeness Given by a Fuzzy Partition

Using the language of fuzzy set theory, we can say that any symmetric fuzzy relation $w \subsetneq X^2$, where symmetry means $\forall x, y \in X \colon w(x, y) = w(y, x)$, is closeness on X.

Note that in the context of fuzzy relations, it holds that for weakly reflexive closeness, we have $\overline{w} \leq 1$. For $\overline{w} = 1$, we obtain a special case called reflexive closeness on X.

Lemma 1 (Closeness Given by a Fuzzy Partition – 1). *Let $\{A_1, \ldots, A_l\}$ be a fuzzy partition associated with nodes $y_1, \ldots, y_l$ of the set $\{x_1, \ldots, x_L\} \subset [y_0, y_{l+1}]$, s.t. $y_0, \ldots, y_{l+1}, x_1, \ldots, x_L \in \mathbb{R}$, $0 < l = L < +\infty$ and $y_0 < y_1 = x_1 < y_2 = x_2 < \cdots < y_l = x_L < y_{l+1}$. For all $i, j = 1, \ldots, l$, let it hold that*

$$A_i(y_j) = A_j(y_i), \tag{3}$$

then the function $w \colon \{x_1, \ldots, x_L\}^2 \to \mathbb{R}$ given by $\forall i, j = 1, \ldots, L \colon w(x_i, x_j) = A_i(x_j)$ is closeness on $\{x_1, \ldots, x_L\}$.

Proof. The function w is obviously real, bivariate and non-negative. Using the property (3), we have $\forall i, j = 1, \ldots, L \colon w(x_i, x_j) = A_i(x_j) = A_i(y_j) = A_j(y_i) = A_j(x_i) = w(x_j, x_i)$, hence w is also symmetric. □

Lemma 1 shows that a fuzzy partition of a subset of a real interval, s.t. its every point is a node and satisfying condition (3), determines closeness on this set.

2.4. Closeness as Weighted Adjacency of Graph Vertices

In this subsection, we show that closeness can be expressed using the language of graph theory. We do not recall the basic concepts of this theory (details can be found, e.g., in [20]).

Generally, closeness between any two objects x_i and x_j in X is described by the values (weights) $w(x_i, x_j)$ of a real, non-negative, symmetric bivariate function w.

Let $G(X, E, W)$ be a weighted graph where the set of vertices corresponds to objects in X, the set of edges $E \subseteq X^2$ and weights of the edges are given by the adjacency matrix W and in accordance with (2). The objects are non-close if and only if the corresponding vertices are not connected by a direct edge, i.e., there is a fictional edge with zero weight between them. We denote this disconnectedness or non-closeness by $\not\sim$ which is a classical (meaning non-fuzzy) symmetric binary relation on X, i.e., $\not\sim \subseteq X^2$:

$$x_i \not\sim x_j \Leftrightarrow w(x_i, x_j) = 0. \tag{4}$$

In the other words, the value of closeness determines the existence of an edge.

By defining the values of closeness $\{w(x_i, x_j) \mid x_i, x_j \in X\}$, i.e., by defining the values of entries of the closeness matrix, we simultaneously define the values of entries of the adjacency matrix W which uniquely determines a weighted graph $G(X, E, W)$ by (1), (2) and (4). That is why the notion of closeness can be also established from the perspective of weighted graphs: let $X = \{x_i \mid i = 1, \ldots, L\}$ be a set of vertices, $L \in \mathbb{N}$, then, following the convention given by (4) as $E \subseteq X^2$, any evaluation of the set of all edges X^2 described by a symmetric, non-negative function $w \colon X^2 \to \mathbb{R}$ is closeness on X.

Below, we join the theory of closeness with a certain type of fuzzy partition to describe the space this paper deals with.

2.5. Initial Assumptions

Let us have a finite set of points on $\mathbb{R}$:

$$X = \{x_i \mid i = 1, \ldots, L\}, \tag{5}$$

and its non-empty subset

$$Y \subseteq X, \quad 0 < |Y| = l \leq L = |X| < +\infty.$$

Let $\{A_t \mid t \in Y\}$ be a fuzzy partition of X with nodes in Y (in the sense of Remark 3), s.t. with each node $t \in Y$, we associate a basic function $A_t \colon X \to [0,1]$. Let, moreover, (3) be satisfied, i.e., $\forall t, s \in Y \colon A_t(s) = A_s(t)$, and let all points in X be covered by at least one basic function, i.e.,

$$\bigcup_{t \in Y} \{x \in X \mid A_t(x) > 0\} = X. \tag{6}$$

Let us define *node degrees* by

$$\forall t \in Y \colon d_{tt} = \sum_{x \in X} A_t(x). \tag{7}$$

Following condition 3 in Definition 6, we know that $A_t(t) > 0$, $t \in Y$, which implies that each node degree $d_{tt} \in (0, \infty)$. Moreover, based on Definition 7, it means that the set Y is sufficiently dense with respect to the fuzzy partition $\{A_t \mid t \in Y\}$; hence, the same holds for the universe X.

It is worth noting that instead of working with the closeness on the whole universe X, given by a function on X^2, we work only with its restriction on $Y \times X$ (the values of closeness on $(X \setminus Y) \times X$ are undefined, as explained in the Introduction). In the other words, closeness $w \colon X^2 \rightharpoonup \mathbb{R}$ is a partial function (partial function $f \colon A \rightharpoonup B$ maps each element of a set A to at most one element of the set B, it is a functional binary relation with carrier set being a subset of $A \times B$) given by

$$\forall x_i, x_j \in X \colon w(x_i, x_j) = \begin{cases} A_t(x_j) & t = x_i \in Y \\ \text{undefined} & \text{otherwise}. \end{cases}$$

This description justifies the selection of nodes within the universe (Semantic description of nodes is also explained in the Introduction).

Even though we will still call the total function $w\colon Y \times X \to \mathbb{R}$ (total function $f\colon A \to B$ maps each element of a set A to exactly one element of the set B, it is a classical function) *closeness on X* and denote the corresponding closeness space as (X, w).

Lemma 2 (Closeness Given by a Fuzzy Partition—2)**.** *Let $\{A_1, \ldots, A_l\}$ be a fuzzy partition associated with nodes $y_1, \ldots, y_l$ of the set $\{x_1, \ldots, x_L\} \subset [y_0, y_{l+1}]$, s.t. $y_0, \ldots, y_{l+1}, x_1, \ldots, x_L \in \mathbb{R}$, $y_0 < y_1 < y_2 < \ldots < y_l < y_{l+1}$, $y_0 < x_1 < x_2 < \ldots < x_L < y_{l+1}$, $\{y_1, \ldots, y_l\} \subseteq \{x_1, \ldots, x_L\}$, $0 < l \leq L < +\infty$ and let condition (3) hold. Let us denote*

$$\forall i = 1, \ldots, L\colon \kappa(i) = \begin{cases} j & x_i = y_j \\ l+1 & \nexists y \in \{y_1, \ldots, y_l\}\colon x_i = y \end{cases}$$

(note that in case $x_i = y_j$, we have $y_j = y_{\kappa(i)}$). Then the function $w\colon \{x_1, \ldots, x_L\}^2 \to \mathbb{R}$ given by $w(x_i, x_j) = A_{\kappa(i)}(x_j)$, where $\kappa(i) \leq l$, is closeness on $\{x_1, \ldots, x_L\}$.

Proof. The function w is obviously real, bivariate and non-negative. Using the property (3), we have $\forall i, j = 1, \ldots, L, \kappa(i), \kappa(j) \leq l\colon w(x_i, x_j) = A_{\kappa(i)}(x_j) = A_{\kappa(i)}(y_{\kappa(j)}) = A_{\kappa(j)}(y_{\kappa(i)}) = A_{\kappa(j)}(x_i) = w(x_j, x_i)$, hence w is also symmetric. □

Lemma 2 shows that a fuzzy partition of a subset of a real interval, s.t. it contains all nodes and satisfies condition (3), determines closeness on this set. We will follow Lemma 2, and hence closeness is given by:

$$w(t, x) = A_t(x), \quad \text{where } t \in Y, x \in X. \tag{8}$$

By that, the closeness space (X, w) is specified. The values of closeness $w\colon Y \times X \to \mathbb{R}$ are inserted in the closeness matrix $W \in \mathbb{R}^{l \times L}$ and given by

$$\forall t \in Y, x \in X\colon W(t, x) = w(t, x). \tag{9}$$

It means that by fixing a set of basic functions, we determine values of the closeness-describing function $w\colon Y \times X \to \mathbb{R}$.

It follows from (4), (6) and (8) that for every point x of the universe X there exists a node $t \in Y$, s.t. $w(t, x) > 0$. It means that every point is connected with at least one node.

Sections 2.1 and 2.2 covered the basic theory of closeness and the theory of fuzzy partitions, respectively. In the following Section 2.6, we recall the theory of fuzzy transform that we need to describe a fundamental property (the F-transform identity) of the space with closeness that will be used in Section 3 to describe the preimage problem.

2.6. Fuzzy Transform

In Section 2.2, we defined the fuzzy partition and its units, basic functions. Below, we proceed by the establishment of fuzzy transform. The closeness space (X, w) serves as a ground on which F-transform is established as its key element, a fuzzy partition, is selected on condition (3) that allows the creation of (X, w) based on (8) in Section 2.5.

Definition 9 (Direct F-Transform [4])**.** *Let X be a set, $Y \subseteq X \subset [y_0, y_{l+1}] \subset \mathbb{R}$, $0 < |Y| = l \leq |X| = L < +\infty$, $\mathcal{P} = \{A_t\colon X \to [0,1] \mid t \in Y\}$ be a fuzzy partition of X with nodes in Y, s.t. $y_0 < x_1 < x_2 < \ldots < x_L < y_{l+1}$, and satisfying (3), i.e., $\forall t, s \in Y\colon A_t(s) = A_s(t)$. Let d_{tt} be a node degree given by (7) and $u\colon X \to \mathbb{R}$ be a function where $\forall i = 1, \ldots, L\colon u(x_i) = u_i$. Then the direct discrete F-transform of u with respect to $\mathcal{P}$ is a vector $F[u] \in \mathbb{R}^l$ of F-transform components defined by*

$$\forall t \in Y \colon F[u]_t = \frac{\sum_{i=1}^{L} u_i A_t(x_i)}{\sum_{i=1}^{L} A_t(x_i)} = \frac{\sum_{i=1}^{L} u_i A_t(x_i)}{d_{tt}}. \tag{10}$$

Remark 4. Note that in [4], the direct F-transform is defined only with respect to a fuzzy partition (without closeness) and in that case, the requirement that $\mathcal{P}$ satisfies (3) is omitted. This paper, however, assumes a closeness background. That is why for every node $t \in Y$, the value

$$d_{tt} = \sum_{i=1}^{L} w_{ti}, \tag{11}$$

is also called a node degree which fully agrees with (7). For the corresponding space with closeness (X, w) where the closeness values are denoted according to (1), we equivalently define the direct F-transform of u as a vector $F[u] \in \mathbb{R}^l$ of F-transform components defined by

$$\forall t \in Y \colon F[u]_t = \frac{\sum_{i=1}^{L} u_i w_{ti}}{\sum_{i=1}^{L} w_{ti}} = \frac{\sum_{i=1}^{L} u_i w_{ti}}{d_{tt}}, \tag{12}$$

which fully agrees with (10).

Let F be a real vector-valued operator that acts on the space of all real functions on the universe X and gives us a real vector in $\mathbb{R}^l$ given by (10) that can be uniquely identified with a function on the set of nodes Y, an element of a functional space on Y, s.t. $\forall t \in Y \colon F[u](t) = F[u]_t$. So,

$$F \colon u \mapsto F[u], \text{ where } u : X \to \mathbb{R} \text{ and } F[u] : Y \to \mathbb{R}.$$

Then Definition 9 describes F-transform as a vector of F-transform components, i.e., as an image of the operator F.

Lemma 3 (F-Transform Identity). *Let X be a set, $Y \subseteq X$, $0 < |Y| = l \leq |X| = L < +\infty$, $\mathcal{P} = \{A_t \colon X \to [0, 1] \mid t \in Y\}$ be a fuzzy partition of X with nodes Y satisfying the assumptions of Lemma 2 and (X, w) be a corresponding space with closeness where $W \in \mathbb{R}^{l \times L}$ is the closeness matrix given by (8) and (9), and $D \in \mathbb{R}^{l \times l}$ be a diagonal scaling matrix with diagonal elements $(d_{tt})_{t \in Y}$ given by (7) and (11). Let $F[u] \in \mathbb{R}^l$ be the vector of F-transform components of the function $u \colon X \to \mathbb{R}$ with respect to $\mathcal{P}$ and $u \in \mathbb{R}^L$ be the vector form of the function $u \colon X \to \mathbb{R}$ given by $\forall i = 1, \ldots, L \colon u_i = u(x_i)$. Then the following holds:*

$$DF[u] = Wu. \tag{13}$$

Equation (13) is called the **F-transform identity**.

Lemma 3 characterizes the operator F in a matrix form utilizing the fact that both functional spaces contain discrete functions; hence, they can be represented as vectors (matrices of finite order):

$$F \colon \mathbb{R}^L \to \mathbb{R}^l, u \mapsto F[u] = D^{-1}Wu.$$

There is a historical reason for representing F-transform as a mapping.

Definition 10 (Inverse F-Transform). *Let X be a set, $Y \subseteq X$, $0 < |Y| = l \leq |X| = L < +\infty$, $\mathcal{P} = \{A_t \colon X \to [0, 1] \mid t \in Y\}$ be a fuzzy partition of X with nodes Y satisfying the assumptions of Lemma 2 and (X, w) be a corresponding space with closeness where $W \in \mathbb{R}^{l \times L}$ is the closeness* matrix given by (8) and (9). Let W^t be its t-th row and $F[u]$ be the direct F-transform of a function $u\colon X \to \mathbb{R}$ in accordance with Definition 9 and (12). Then the inverse F-transform of the function u is the vector

$$\hat{F}[u] = W^\top F[u] = \sum_{t \in Y} F[u]_t {W^t}^\top \in \mathbb{R}^L\,.$$

Definition 10 agrees with the standard case given in [4] where it is shown that the inverse F-transform approximates the original function defined on the fuzzy partitioned space (the function is sparsely represented by the direct F-transform). Note that it is called "inverse" as it transforms the vector of direct F-transform components back to the original space. It is not an inverse mapping, though.

3. Preimage Problem

In this section, we describe a form of preimage problem (see Definition 11 below) that we would like to solve in order to describe a similarity between functions. The inverse F-transform of a function u provides a vector $\hat{F}[u]$ that approximates the vector form of u. Therefore, we were motivated to solve the problem of how to find the class of functions that have the same inverse F-transform but decided to solve a more specific problem described in the following paragraph.

Let us have a fixed mapping (based on the F-transform identity in Lemma 3) that maps functions on vectors: $u \mapsto DF[u] = Wu$. We would like to characterize the similarity between functions in the sense that two functions are similar if their images (scaled vectors of direct F-transform components) coincide. In that case, their inverse F-transforms coincide as well ($DF[u'] = DF[\overline{u}] \Rightarrow \hat{F}[u'] = \hat{F}[\overline{u}]$). We use the singular value decomposition (formalized, e.g., in [21] and commonly abbreviated as SVD) of the closeness matrix W to specify the problem and its solution.

We assume that the universe X, its subset Y of nodes and the closeness matrix $W \in \mathbb{R}^{l \times L}$ given by (8) and (9), where $0 < |Y| = l \leq |X| = L < +\infty$, are fixed. Recall that the node degrees d_{tt}'s are given by (7) and form the diagonal of the matrix $D \in \mathbb{R}^{l \times l}$. In the following text, we will respect the aforegiven assumptions.

Moreover, for simplicity, we will use the language of linear algebra in the following two subsections. That is why the clarification of the correspondence between discrete functions (functional spaces) and vectors (vector spaces) follows.

From the functional viewpoint, let A be the space of all functions $u\colon X \to \mathbb{R}$ and let B be the space of such functions $f\colon Y \to \mathbb{R}$ that assign to each node $t \in Y$ the value of the corresponding F-transform component multiplied by the corresponding node degree, i.e.,

$$f \in B \quad \Leftrightarrow \quad \exists u\colon X \to \mathbb{R}\, \forall t \in Y\colon f(t) = d_{tt} F[u]_t\,.$$

From the viewpoint of the linear algebra, recalling that $|X| = L$, $A = \mathbb{R}^L$ is a vector space, its elements are later denoted by u or v. Space A contains all vector representations of functions on X. As each element of A corresponds to one function on the universe X, we can further identify X, given by (5), with the set of its indices $\{1, \ldots, L\}$. Hence, the correspondence between the function $u\colon X \to \mathbb{R}$ and the vector $u \in \mathbb{R}^l$ is given by the following equation:

$$\forall x_i \in X\colon u(x_i) = u_i\,. \tag{14}$$

Following (8) and (9), recall that each row of the closeness matrix $W \in \mathbb{R}^{l \times L}$ defines one basic function. Then, based on the F-transform identity, $B = \mathrm{range}\, W$ (range, or column space, of the matrix $W \in \mathbb{R}^{l \times L}$ is a linear subspace of $\mathbb{R}^l$ spanned by all columns of W) is a vector subspace of $\mathbb{R}^l$ containing the scaled images of the operator F that are uniquely determined by W. In the other words, the set of all vectors (elements) of B, $f \in \mathbb{R}^l$, is defined by

$$f \in B \quad \Leftrightarrow \quad \exists u \in A\colon f = DF[u] = Wu\,.$$

Hence, the correspondence between the function $f\colon Y \to \mathbb{R}$ (satisfying $\forall t \in Y\colon f(t) = d_{tt}F[u]_t$) and the vector $f \in \operatorname{range} W$ is given by the following equation:

$$\forall t \in Y\colon f(t) = f_t. \tag{15}$$

Note that, for convenience, the vectors in $\mathbb{R}^l$ are indexed based on the indices of nodes in $\mathbb{R}^L$, so their indices need not form an arithmetic sequence. The same convention will be also respected for the rows of the matrix W.

In the following text, we will also respect the notation of the vector spaces A and B.

3.1. Problem Formulation

In this subsection, we describe the preimage problem in terms of a mapping between two linear vector spaces.

Definition 11 (Preimage Problem). *Let (X, w) be a closeness space with a closeness matrix $W \in \mathbb{R}^{l \times L}$ given by $\forall t \in Y, x \in X\colon w(t, x) = W(t, x)$, where Y is a non-empty subset of X and $0 < l \leq L < +\infty$. Let $A = \mathbb{R}^L$, $B = \operatorname{range} W \subseteq \mathbb{R}^l$ and let an element $b \in B$ be given. Let the direct mapping $W\colon A \to B$ be given by the closeness matrix W, i.e., $\forall v \in A\colon W(v) = Wv$. Then the* preimage problem *is to find the set $\{v \in A \mid Wv = b\}$. This set is the* preimage *of b.*

Although the solution to the preimage problem is a set, we will call any of its elements a *solution to the preimage problem*. By choosing the element $b \in B$, we want to utilize the closeness on X to induce an equivalence relation on A to describe the similarity between vectors.

The closeness matrix W generally exists for any closeness (based on (1) and (2)) on a finite set. It can be also defined for the special case that originates in a certain fuzzy partition (based on (8) and (9)). Hence, the preimage problem described in Definition 11 is a general formulation of the problem. By solving the general one (described by the closeness matrix), we simultaneously solve the problem specified by F-transform.

Assume that we found an SVD (see [21] to recall this technique) of the closeness matrix W in the form of the product of three matrices $PSZ^\top$, where $P \in \mathbb{R}^{l \times l}$ and $Z \in \mathbb{R}^{L \times L}$ are orthogonal (these matrices are not determined uniquely) and $S \in \mathbb{R}^{l \times L}$ is diagonal ($\forall i, j\colon i \neq j \Rightarrow s_{ij} = 0$) with so-called singular values $\forall i = 1, \ldots, l\colon \sigma_i = s_{ii}$ on its diagonal, s.t. $\sigma_1 \geq \sigma_2 \geq \cdots \geq \sigma_k > \sigma_{k+1} = \cdots = \sigma_l = 0$ where $k = \dim(\operatorname{range} W) = \operatorname{rank} W$ is the number of linearly independent rows of W ($0 < k \leq l$). Any complex-valued matrix can be represented in an SVD form where the singular values are invariant under different choices of P and Z.

Lemma 4. *Let $W \in \mathbb{R}^{l \times L}$ be a closeness matrix and $PSZ^\top$ be its SVD as above, then*

$$\operatorname{range} W = \operatorname{range} PSZ^\top = \left\{ \sum_{i=1}^{k} c_i p_i \,\middle|\, c_i \in \mathbb{R} \right\},$$

where p_i is the i-th column of the matrix P.

Lemma 4 states that $\operatorname{range} W$ is spanned by the left singular vectors (p_i, $i = 1, \ldots, k$) corresponding to positive singular values of W.

Proof of Lemma 4. Let $y \in \operatorname{range} W$, i.e.,

$$y = \sum_{j=1}^{L} d_j W_j,$$

where for $j = 1, \ldots, L$, $d_j \in \mathbb{R}$ and W_j is the j-th column of W. For any $r = 1, \ldots, l$, we have

$$y_r = \sum_{j=1}^{L} d_j w_{rj} = \sum_{j=1}^{L} d_j (PSZ^\top)_{rj} = \sum_{j=1}^{L} d_j \sum_{i=1}^{k} p_{ri} \sigma_i z_{ji} = \sum_{i=1}^{k} p_{ri} \sum_{j=1}^{L} d_j \sigma_i z_{ji}.$$

Let

$$\mathcal{S} = \left\{ \sum_{i=1}^{k} c_i p_i \,\bigg|\, c_i \in \mathbb{R} \right\}.$$

If for each $r = 1, \ldots, k$, we set

$$c_i = \sum_{j=1}^{L} d_j \sigma_i z_{ji},$$

this proves that $y \in \mathcal{S}$ because the value c_i does not depend on r. Hence, range $W \subseteq \mathcal{S}$. As both of these sets are k-dimensional subspaces of $\mathbb{R}^l$, they must coincide. □

3.2. Problem Solution

Firstly, we discuss conditions when the preimage problem has a solution and when this solution is unique.

As $Y \neq \varnothing$, the matrix W has a positive number of rows and, hence, a positive number of columns, so range $W \neq \varnothing$ and a vector b can be always selected. Following Definition 11, we were given a vector $b \in B$. This means that the solution always exists. The solution is always unique—it is the set of all vectors $v \in A$ that are mapped on b.

Let us now discuss when the solution to the preimage problem is formed by a set with exactly one element. Following the well-known Fredholm alternative, if the homogeneous system $Wv_0 = 0$ has a nontrivial solution $v_0 \in A$, then the set has more than one element. If $Wv_0 = 0$ has only the trivial solution $v_0 = 0$, then there is only one vector v solving the problem. Details can be found in Section 3.2.3.

Secondly, we characterize the solution from three perspectives: the first one is a useful characterization (Lemmas 5 and 6), the latter two (Theorem 1 together with Corollary 1, and Lemma 9) describe what the solution looks like.

3.2.1. Weighted Arithmetic Mean

Definition 12 (Weighted Arithmetic Mean of a Vector)**.** *With respect to the real weights $(w_i)_{i=1}^{L}$ that satisfy $\sum_{i=1}^{L} w_i \neq 0$, the weighted arithmetic mean of a vector $[v_1, \ldots, v_L]^\top \in \mathbb{R}^L$ is a real number*

$$\frac{\sum_{i=1}^{L} v_i w_i}{\sum_{i=1}^{L} w_i}.$$

The equation of the preimage problem, $Wv = b$, is a set of linear equations $W^t v = b_t$ where W^t is a row of the matrix W corresponding to the index $t \in Y$, and we solve

$$b_t = \sum_{i=1}^{L} v_i w_{ti} \quad \text{where } t \in Y. \tag{16}$$

We have $l = |Y|$ equations in the form (16) of $L \geq l$ variables $v_1, \ldots, v_L \in \mathbb{R}$, from which stems the following lemma.

Lemma 5. *Any vector $v \in A$ is a solution to the preimage problem $Wv = b$ if and only if for every index $t \in Y$, its weighted arithmetic mean with respect to closeness-defining weights $(w_{ti})_{i=1}^{L}$ is equal to $\frac{b_t}{d_{tt}}$, where b_t is the t-th component of the vector b and d_{tt} is the node degree given by* (11).

Proof. Dividing both sides of (16) by $d_{tt} = \sum_{i=1}^{L} w_{ti}$, the claim is obvious. □

Lemma 5 states that with respect to every row of W, the weighted arithmetic means of the solutions to the preimage problem are equal. By that, Lemma 5 characterizes the set of all solutions to the preimage problem $Wv = b$ for a given vector $b \in B$. Hence, the similarity we are looking for is determined by the weighted mean of vectors. Moreover, it demarcates the solution as the equivalence class of vectors (discrete functions):

Lemma 6. *Let us have the equivalence relation on A given by: for all $v', \overline{v} \in A$, we have*

$$v' \equiv \overline{v} \Leftrightarrow \forall t \in Y: \frac{\sum_{i=1}^{L} v'_i w_{ti}}{\sum_{i=1}^{L} w_{ti}} = \frac{\sum_{i=1}^{L} \overline{v}_i w_{ti}}{\sum_{i=1}^{L} w_{ti}}.$$

Let $b = Wu$ for some $u \in A$ (i.e., $b \in B$), then the solution to the preimage problem $Wv = b$ is the set $\{v \in A \mid v \equiv u\}$.

Proof. This lemma is a direct consequence of Lemma 5 where for any $v \equiv u$, we have

$$\forall t \in Y: \frac{\sum_{i=1}^{L} v_i w_{ti}}{\sum_{i=1}^{L} w_{ti}} = \frac{\sum_{i=1}^{L} u_i w_{ti}}{\sum_{i=1}^{L} w_{ti}} = \frac{b_t}{d_{tt}},$$

and because u is a solution to the preimage problem, any $v \equiv u$ is a solution.

Conversely, consider a vector $v^* \not\equiv u$, s.t. $Wv^* = b$. Then

$$\exists t^* \in Y: \frac{\sum_{i=1}^{L} v_i^* w_{ti}}{\sum_{i=1}^{L} w_{ti}} \neq \frac{\sum_{i=1}^{L} u_i w_{ti}}{\sum_{i=1}^{L} w_{ti}} = \frac{b_t}{d_{tt}},$$

which, based on (16), contradicts $Wv^* = b$. This contradiction proves that the equivalence class contains exactly all solutions to the preimage problem. □

3.2.2. Pseudoinverses

Decomposing W into its SVD by $W = PSZ^\top$ gives the preimage problem in the form $PSZ^\top v = b$ where $P \in \mathbb{R}^{l \times l}$ and $Z \in \mathbb{R}^{L \times L}$ are orthogonal and $S \in \mathbb{R}^{l \times L}$ is diagonal (i.e., $\forall i \neq j: s_{ij} = 0$) with singular values of W, $\sigma_t = s_{tt}$, $t \in Y$, on its diagonal. The equation at hand is thus

$$\forall t \in Y: \sum_{q \in X} \sum_{r \in Y} \sigma_r p_{tr} z^\top_{tq} v_q = b_t,$$

where $z^\top_{tq} = z_{qt}$ is the element of the matrix $Z^\top$ corresponding to the index t in its q-th column.

Theorem 1 (Explicit Solution). *Let $W \in \mathbb{R}^{l \times L}$ be the closeness matrix given by (8) and (9), its SVD $W = PSZ^\top$ be given by matrices $P \in \mathbb{R}^{l \times l}$, $S \in \mathbb{R}^{l \times L}$ and $Z \in \mathbb{R}^{L \times L}$, s.t. rows of P, columns of P and rows of S are indexed by Y, $Y^+ = \{t \in Y \mid \sigma_t > 0\}$. Let, moreover, $b \in B$. Then, an explicit solution to the matrix equation $Wv = b$ with respect to the given SVD is the vector*

$$v = \sum_{t \in Y^+} \frac{b^\top p_t}{\sigma_t} z_t \in A,$$

where p_t is the t-th column of the matrix P and z_t is the t-th column of the matrix Z.

Proof. Let 1_t denote the vector containing all zeros but 1 on the t-th coordinate and $(\sigma_t^{-1})_t$ denote the vector containing all zeros but σ_t^{-1} on the t-th coordinate. Following the orthogonality of P, we have:

$$Wv = PSZ^\top \sum_{t \in Y^+} \frac{b^\top p_t}{\sigma_t} z_t = \sum_{t \in Y^+} \frac{b^\top p_t}{\sigma_t} PSZ^\top z_t = \sum_{t \in Y^+} \frac{b^\top p_t}{\sigma_t} PS1_t$$
$$= \sum_{t \in Y^+} b^\top p_t PS(\sigma_t^{-1})_t = \sum_{t \in Y^+} b^\top p_t P1_t = \sum_{t \in Y^+} b^\top p_t p_t = b.$$

□

Remark 5. *Note that the matrices P and Z in Theorem 1 are not determined uniquely, and so there is generally more than one vector $v \in A$ solving $Wv = b$. Following Lemma 6, the solution to the corresponding preimage problem is given by the set $\{v' \in A \mid v' \equiv v\}$ formed by all vectors equivalent with the explicit solution v.*

Another general description of the solution is expressed using the notion of right inverse matrix, also called (right) pseudoinverse. Any matrix $W^{-1} \in \mathbb{R}^{L \times l}$ satisfying $WW^{-1} = I$ where $I \in \mathbb{R}^{l \times l}$ is the identity matrix, is the *right inverse matrix* of $W \in \mathbb{R}^{l \times L}$; if we demanded that $W^{-1}W$ were symmetric, W^{-1} would be unique, otherwise there can be multiple right inverse matrices.

Lemma 7. *Let $W^{-1} \in \mathbb{R}^{L \times l}$ be any right inverse matrix of the closeness matrix W and $b \in B$, then the vector $v = W^{-1}b$ is a solution to the preimage problem $Wv = b$ with respect to the given right inverse matrix W^{-1}.*

Proof. $Wv = WW^{-1}b = Ib = b$. □

The following corollary is a special case of Theorem 1 written in a matrix form.

Corollary 1. *Let $W \in \mathbb{R}^{l \times L}$ be the closeness matrix and $\operatorname{rank} W = l$. If $W = PSZ^\top$ is its fixed SVD, then its right inverse is obtained as $W^{-1} = ZS^+P^\top$ where $S^+ \in \mathbb{R}^{L \times l}$ is the right inverse of S, i.e., a diagonal matrix with inverted singular values of S on its diagonal. Let $b \in B$, then $v = ZS^+P^\top b$ is a solution to the preimage problem $Wv = b$.*

Proof. Following the orthogonality of matrices P and Z, we have $WW^{-1} = WZS^+P^\top = PSZ^\top ZS^+P^\top = PSS^+P^\top = PP^\top = I$, hence, $Wv = WW^{-1}b = b$. □

Lemma 8. *If a matrix $W \in \mathbb{R}^{l \times L}$, $0 < l \leq L$, has linearly independent rows, then $W^{-1} = W^\top(WW^\top)^{-1}$ is its right inverse.*

Proof. Following SVD of W used in the Corollary 1, full rank of W ensures the existence of $(WW^\top)^{-1}$. Then, $WW^{-1} = WW^\top(WW^\top)^{-1} = I$. □

Note that the condition that no singular value of W is zero is equivalent to the condition in Lemma 8 that W has linearly independent rows.

Example A1 in the Appendix A illustrates the preimage problem and shows one vector that solves it.

3.2.3. Affine Subspace

The solution to the preimage problem does not form a linear subspace because, e.g., $W(v' + \overline{v}) = 2b$ which is generally not equal to b. It is a linear subspace of A if and only if $b = 0$, i.e., if and only if it is equal to null W (right nullspace, or kernel, of a matrix $W \in \mathbb{R}^{l \times L}$ contains all vectors $a \in \mathbb{R}^L$, s.t. $Wa = 0$, it is a linear subspace of $\mathbb{R}^L$). In the other words, if $b = 0$, then the preimage problem, $Wv = b$, becomes a homogeneous system of linear equations, making its solution be a linear subspace of A.

Coming back to the general case where the vector $b \in B$ is arbitrary, we obtain that the solution to the preimage problem forms an affine subspace of A.

Lemma 9. *Let $b \in B$ and $v' \in A$ be a particular solution to the preimage problem $Wv = b$, then the set of all vectors that solve this problem, forms an affine subspace $\{v' + v_0 \mid v_0 \in \text{null } W\}$ of A.*

Proof. For the vector v', it holds that $Wv' = b$ and for any vector $v_0 \in \text{null } W$, it holds that $Wv_0 = 0$. Hence, for any element $a = v' + v_0$ of the set $\mathfrak{A} = \{v' + v_0 \mid v_0 \in \text{null } W\}$, it holds that $Wa = Wv' + Wv_0 = b + 0 = b$, so every element of $\mathfrak{A}$ is a solution to the preimage problem.

Conversely, consider a vector $a' \in A$, s.t. $Wa' = b$ and $a' \notin \mathfrak{A}$. Then $W(a' - v') = Wa' - Wv' = b - b = 0$, so the vector $a' - v' \in \text{null } W$. As the vector a' can be expressed as $v' + a' - v'$, we see that $a' \in \mathfrak{A}$. This contradiction proves that the set $\mathfrak{A}$ contains exactly all solutions to the preimage problem. Moreover, it proves that the set $\mathfrak{A}$ is unique, i.e., if we express the solution to the preimage problem as $\mathfrak{B} = \{\bar{v} + \bar{v_0} \mid \bar{v_0} \in \text{null } W\}$ for any vector $\bar{v} \in A$, s.t. $W\bar{v} = b$, then $\mathfrak{A} = \mathfrak{B}$.

As null $W \subseteq\subseteq A$, $\mathfrak{A}$ is an affine subspace of A with the displacement vector v'. □

Using the SVD of W as above, we can express the solution to the preimage problem by right singular vectors corresponding to zero singular values of W as

$$\left\{ v' + \sum_{i=k+1}^{L} c_i z_i \,\middle|\, c_i \in \mathbb{R} \right\}.$$

The dimension of that affine subspace is then $\dim(\text{null } W)$ which is zero (and hence exists only one vector solving the preimage problem) if and only if $k = l = L$, i.e., the closeness matrix W is square and regular. If W is rectangular, we have infinitely many vectors $a \in A$ solving the preimage problem.

Corollary 2. *Any weighted mean of elements of the set of all vectors solving the preimage problem with respect to weights with nonzero sum is again an element of this set.*

Proof. Let $n \in \mathbb{N}$ and $\forall i = 1, \ldots, n$: $Wv^i = b$, i.e., each $v^i \in A$ be a solution to the preimage problem. Let $c_1, \ldots, c_n \in \mathbb{R}$, s.t. $\sum_{i=1}^{n} c_i \neq 0$. Then it holds that

$$W\left(\frac{\sum_{i=1}^{n} c_i v^i}{\sum_{i=1}^{n} c_i} \right) = \frac{\sum_{i=1}^{n} c_i W v^i}{\sum_{i=1}^{n} c_i} = \frac{\left(\sum_{i=1}^{n} c_i\right) b}{\sum_{i=1}^{n} c_i} = b,$$

which proves the claim. □

In this subsection, we described the solution to the preimage problem from three different perspectives: using weighted means of functional values (determining an equivalence class of mutually similar functions), using right inverses of the closeness matrix (that can be obtained also by SVD) and using the notion of affine subspace (its displacement vectors can be also obtained by SVD of W, based on Theorem 1).

In the next section, we show the connection between the inverse F-transform and a solution to the preimage problem as the rows of W are formed by the basic functions of the fuzzy partition.

4. Inverse F-Transform

In this section, we omit the assumption (6) stating every point of the universe X must be covered by at least one basic function. Admitting this degenerated case allows us to express a useful characterization of the case when the inverse F-transform, given by Definition 10, provides a solution to the preimage problem.

Lemma 10. *Let the assumptions of Lemma 3 be satisfied. If*

$$WW^\top = D, \quad (17)$$

then the inverse F-transform $v = \hat{F}[u]$ given by Definition 10 is a solution to the preimage problem according to Definition 11, $DF[v] = Wv = Wu = DF[u] = b$ for any vector $u \in A$.

Proof. We see that the inverse F-transform is a solution to the preimage problem if and only if for any $u \in A$, it holds:

$$Wu = DF[u] = b \quad = \quad Wv = W\hat{F}[u] = WW^\top F[u] = WW^\top D^{-1} Wu,$$

which holds if

$$WW^\top D^{-1} = I,$$

or equivalently if (17) holds true. □

Corollary 3. *Condition (17) holds true if both of the following equations are fulfilled:*

1. $\forall t \in Y \colon d_{tt} = \sum_{i=1}^L w_{ti} \quad = \quad \sum_{i=1}^L w_{ti}^2,$
2. $\forall t, s \in Y, t \neq s \colon d_{ts} = 0 \quad = \quad \sum_{i=1}^L w_{ti} w_{si}.$

Recalling that $w_{ti} = W(t, x_i) = A_t(x_i) \in [0,1]$, we infer that condition 1 is equivalent to stating that $\forall t \in Y, x_i \in X \colon w_{ti} \in \{0,1\}$, i.e., only simple-minded assignment of closeness is possible to be used. Condition 2 is equivalent to stating that every column of the matrix W contains at most one non-zero value—let us denote the system of all such matrices as $\mathfrak{W}$. Putting these two conditions together, we obtain the following corollary.

Corollary 4. *If the matrix $W \in \mathfrak{W}$ contains only values 0 and 1, then the inverse F-transform $v = \hat{F}[u]$ given by Definition 10 is a solution to the preimage problem.*

This property is demonstrated in Examples A2 and A3 in the Appendix A.

5. New Set of Basic Functions

This section was inspired by the paper [17] where one of the main goals consists in finding conditions under which a noisy signal must be sampled so that it can be reconstructed from its F-transform components. The authors showed that a fuzzy partition adjoint to the partition (with the same nodes) that determined the F-transform components, ensures that the associated inverse F-transform provides the best approximation of the original, continuous function in a certain space.

We analyzed a possible connection between discrete fuzzy partitions on the closeness space and decided to address the following question: can we find a solution to the preimage problem $Wv = b$ by creating a new set of basic functions, s.t. they determine a linear transformation of the vector b that would solve the problem? If so, what are the conditions that this new set of basic functions must satisfy?

Throughout this section, assume that the sets $X = \{1, \ldots, L\}$, Y and $\{A_t \colon X \to [0,1] \mid t \in Y\}$, the matrices W and D and the vector b are related as in Lemma 3 and fixed.

We are looking for a new set of basic functions (a new fuzzy partition) $\{B_t: X \to \mathbb{R} \mid t \in Y\}$, written in rows of a newly created matrix $M \in \mathbb{R}^{l \times L}$, i.e.,

$$\forall t \in Y, x \in X: m_{tx} = B_t(x), \qquad (18)$$

s.t. the vector $v = M^\top b$ solves the preimage problem.

Therefore, we require that $WM^\top b = b$ and from this, we deduce properties of the matrix M and express them in terms of basic functions A_t's and B_t's. Let for an arbitrary matrix N, N^t denote its t-th row, N_t denote its t-th column and N_{ts} denote its entry n_{ts}.

Hence, we require that

$$\forall t \in Y: (WM^\top)^t b = \sum_{s \in Y} (WM^\top)_{ts} b_s = \sum_{s \in Y} W^t (M^\top)_s b_s = b_t.$$

Recalling that $W_{tx} = A_t(x)$ and $(M^\top)_{xs} = M_{sx} = B_s(x)$, for $t, s \in Y, x \in X$, we have

$$\forall t \in Y: \sum_{s \in Y} \left(b_s \sum_{x \in X} A_t(x) B_s(x) \right) = b_t. \qquad (19)$$

Substituting the constants

$$c_{ts} = \sum_{x \in X} A_t(x) B_s(x), \quad (t, s) \in Y \times Y,$$

in (19), we obtain that

$$\forall t \in Y: \sum_{s \in Y} b_s c_{ts} = b_t.$$

In the other words, by creating the matrix $C = WM^\top \in \mathbb{R}^{l \times l}$ from the constants c_{ts}, $(t, s) \in Y \times Y$, we solve $Cb = b$. As B can be the whole space $\mathbb{R}^l$, we generally demand that C is an identity matrix I (in the special case of rank $W < l$, this requirement would be unnecessarily strong but leads again to a good choice of new basic functions B_t's).

This means that for all $(t, s) \in Y \times Y$, we demand

1. $\forall t = s: c_{ts} = \sum_{x \in X} A_t(x) B_s(x) = 1$, and
2. $\forall t \neq s: c_{ts} = \sum_{x \in X} A_t(x) B_s(x) = 0$.

This observation motivates the following definition.

Definition 13 (Compatible Set of Basic Functions). *Let the assumptions of Lemma 3 be satisfied. We say that the set $\{B_t: X \to \mathbb{R} \mid t \in Y\}$ is a compatible set of basic functions with respect to $\{A_t: X \to \mathbb{R} \mid t \in Y\}$ in the closeness space (X, w), if*

$$\forall t \in Y: \sum_{x \in X} A_t(x) B_t(x) = 1,$$

and

$$\forall t, s \in Y, t \neq s: \sum_{x \in X} A_t(x) B_s(x) = 0.$$

Theorem 2 (Solution Determined by a Compatible Set of Basic Functions). *Let the assumptions of Lemma 3 be satisfied and let $\{B_t: X \to \mathbb{R} \mid t \in Y\}$ be a compatible set of basic functions with respect to $\{A_t: X \to \mathbb{R} \mid t \in Y\}$ in the closeness space (X, w). Let $b \in B$ and a matrix $M \in \mathbb{R}^{l \times L}$ be defined by $M_{tx} = B_t(x), t \in Y, x \in X$, then the vector $v = M^\top b$ is a solution to the preimage problem $Wv = b$.*

Proof. $Wv = WM^\top b \in \mathbb{R}^l$ and for each of its components, we have

$$\forall t \in Y: W^t v = (WM^\top)^t b = \sum_{s \in Y} (WM^\top)_{ts} b_s = \sum_{s \in Y} W^t (M^\top)_s b_s$$
$$= \sum_{s \in Y} \sum_{x \in X} W_{tx} M_{sx} b_s = \sum_{s \in Y} \sum_{x \in X} A_t(x) B_s(x) b_s = b_t.$$

This proves that $Wv = b$. □

If we can find a set of new basic functions B_t's compatible with A_t's, then the vector $v = M^\top b$ can be described as an inverse F-transform of an unknown vector $u \in A$, s.t. $b = DF[u]$, with respect to $\{B_t : X \to \mathbb{R} \mid t \in Y\}$ written in the rows of M.

In other words, the direct mapping between the vector spaces A and B is given by W ($W : A \to B$) and any $M^\top$ gives an element of the induced inverse relation, i.e., $(b, M^\top b) \in W^{-1}$ where $W^{-1} \subseteq B \times A$ is the inverse relation to the mapping W. This leads to multiple solutions forming a subset of the affine subspace described in Lemma 9.

Lemma 11. *Let the assumptions of Lemma 3 be satisfied. If rank $W = l$, then there always exists a compatible set of basic functions with respect to $\{A_t : X \to \mathbb{R} \mid t \in Y\}$ in the closeness space (X, w).*

Proof. Following Lemma 8, $W^{-1} = W^\top (WW^\top)^{-1}$ is a right inverse to W. By setting $M^\top = W^{-1}$, we found a compatible set of basic functions $\{B_t : X \to \mathbb{R} \mid t \in Y\}$ given by (18). □

To conclude, we found the inverse-F-transform-like procedure associated with the system of new basic functions that, applied on b, gives a solution to the preimage problem. Examples A4 and A5 in the Appendix A illustrate the proposition of this section.

6. Conclusions

In this paper, we discussed the preimage problem in the space where the relationship between objects is determined by closeness. We showed that any metric can be transformed into closeness and therefore, the latter is weaker than the former. We interlaced closeness with fuzzy partition and characterized both by the closeness matrix.

We expressed similarity between functions based on their images (coinciding with vectors of scaled F-transform components) computed using the closeness matrix. By that (and by setting the basic structure of the space without metric, or norm), we contributed to the mathematical theory in the field of functional analysis. We formulated the preimage problem using the language of matrix calculus. The preimage problem solution is given by (i) a weighted arithmetic mean, (ii) any right inverse of the closeness matrix or (iii) any element of a certain affine subspace. Singular value decomposition was applied to describe the problem and its solution.

We defined the notion of compatible set of basic functions and found conditions under which the inverse F-transform with respect to the given and compatible set of basic functions forms a solution to the preimage problem.

Theoretical results were illustrated by numerical examples. They demonstrate, e.g., that requiring reflexivity of closeness can be counterproductive.

The future research will be focused on imposing further conditions on both collections of basic functions (A_t's and B_t's) to reveal stronger connections between the spaces A and B.

Author Contributions: Investigation, J.J. and I.P.; resources, J.J. and I.P.; writing—original draft preparation, J.J.; writing—review and editing, J.J. and I.P.; supervision, I.P. All authors have read and agreed to the published version of the manuscript.

Funding: The support of the project SGS20/PřF-MF/2022 of the University of Ostrava is greatly appreciated.

Institutional Review Board Statement: Not applicable.

Informed Consent Statement: Not applicable.

Data Availability Statement: Not applicable.

Conflicts of Interest: The authors declare no conflict of interest.

Symbols and Abbreviations

The following symbols and abbreviations are used in this manuscript:

w	closeness, closeness-describing bivariate weight function
(X, w)	closeness space with the carrier set (universe) X and closeness w
$\times$	Cartesian product
W	closeness matrix storing weights $w_{ij} = w(x_i, x_j)$ describing closeness between x_i and x_j
$\subsetsim$	fuzzy subset of a given universe
A_t	basic function associated with the node $t \in Y$
$l = |Y|$	number of nodes within the universe X
$L = |X|$	number of all points of the universe X
d_{tt}	degree of the node $t \in Y$, diagonal element of the scaling (degree) matrix D
$F[u] = [F[u]_t]_{t \in Y}$	F-transform of the function u, vector of F-transform components
$\hat{F}[u]$	inverse F-transform of the function u
$\cdot^\top$	transpose of $\cdot$
W^t	t-indexed row of the closeness matrix W
A	space of all real functions on the universe X or all corresponding L-dimensional vectors
$B = \text{range } W$	space of all real l-dimensional vectors that belong to the column space of the closeness matrix W
P, S, Z	SVD matrices: m. of left singular vectors, m. with singular values σ_t on its diagonal, m. of right singular vectors, respectively
$\equiv$	equivalence relation between vectors in A
W^{-1}	right inverse matrix of the closeness matrix W
$\mathfrak{W}$	system of all real $l \times L$ matrices containing at most one non-zero entry in each column
F-transform	fuzzy transform
SVD	singular value decomposition
s.t.	such that

Appendix A

This section contains numerical examples related to the previous five sections (excluding Section 6) to increase their legibility.

Example A1 relates to Section 3.2, explains the notations of matrices and shows the detailed computation of the F-transform components which will be omitted in other examples. It illustrates the preimage problem and shows one vector that solves it.

Example A1. *Let a set of nodes $Y = \{3, 4, 6\}$ be a selected subset of $X = \{1, 2, 3, 4, 5, 6, 7, 8\}$, the universe, and*

$$\forall t \in Y \colon A_t(x) = \begin{cases} \frac{x-t}{3} + 1 & x \in [t-3, t] \cap X \\ \frac{t-x}{3} + 1 & x \in [t, t+3] \cap X \\ 0 & x \in X \setminus [t-3, t+3] \end{cases}$$

be triangular basic functions on X associated with elements of Y that determine the closeness-describing (weight) matrix

$$W = \begin{bmatrix} \frac{1}{3} & \frac{2}{3} & 1 & \frac{2}{3} & \frac{1}{3} & 0 & 0 & 0 \\ 0 & \frac{1}{3} & \frac{2}{3} & 1 & \frac{2}{3} & \frac{1}{3} & 0 & 0 \\ 0 & 0 & 0 & \frac{1}{3} & \frac{2}{3} & 1 & \frac{2}{3} & \frac{1}{3} \end{bmatrix},$$

and the scaling (node degree) matrix

$$D = \begin{bmatrix} 3 & 0 & 0 \\ 0 & 3 & 0 \\ 0 & 0 & 3 \end{bmatrix}.$$

Note that these two matrices, W and the derived D, are independent of the functions acting on X. Let $u = x^2$ be a particular function on X which has a vector form

$$u = \begin{bmatrix} 1 & 4 & 9 & 16 & 25 & 36 & 49 & 64 \end{bmatrix}^\top.$$

The function u is described by its values $(u_i)_{i=1}^L = (u(x_i))_{x_i \in X}$ that represent registered values of a certain variable (physical, chemical, etc.) measured at all points (on all objects) of the universe (data set) X. Another possible interpretation of u consists in assuming that there is a continuous signal that spreads over a certain medium during a time frame described by an interval $[y_0, y_{l+1}] \supset X$ and we are able to register that signal only in discrete time steps $(x_i)_{i=1}^L$. The components of the direct F-transform of the function u corresponding to all A_t's ($t \in Y$) are:

$$F[u]_1 = \left(1 \cdot \frac{1}{3} + 4 \cdot \frac{2}{3} + 9 \cdot 1 + 16 \cdot \frac{2}{3} + 25 \cdot \frac{1}{3}\right) : \left(\frac{1}{3} + \frac{2}{3} + 1 + \frac{2}{3} + \frac{1}{3}\right) = \frac{31}{3}$$

$$F[u]_2 = \left(4 \cdot \frac{1}{3} + 9 \cdot \frac{2}{3} + 16 \cdot 1 + 25 \cdot \frac{2}{3} + 36 \cdot \frac{1}{3}\right) : \left(\frac{1}{3} + \frac{2}{3} + 1 + \frac{2}{3} + \frac{1}{3}\right) = \frac{52}{3}$$

$$F[u]_3 = \left(16 \cdot \frac{1}{3} + 25 \cdot \frac{2}{3} + 36 \cdot 1 + 49 \cdot \frac{2}{3} + 64 \cdot \frac{1}{3}\right) : \left(\frac{1}{3} + \frac{2}{3} + 1 + \frac{2}{3} + \frac{1}{3}\right) = \frac{112}{3}.$$

Writing them in the vector form, we obtain

$$F[u] = \begin{bmatrix} \frac{31}{3} & \frac{52}{3} & \frac{112}{3} \end{bmatrix}^\top.$$

Obviously, it holds that $DF[u] = Wu$, i.e.,

$$\begin{bmatrix} 3 & 0 & 0 \\ 0 & 3 & 0 \\ 0 & 0 & 3 \end{bmatrix} \cdot \begin{bmatrix} \frac{31}{3} \\ \frac{52}{3} \\ \frac{112}{3} \end{bmatrix} = \begin{bmatrix} \frac{1}{3} & \frac{2}{3} & 1 & \frac{2}{3} & \frac{1}{3} & 0 & 0 & 0 \\ 0 & \frac{1}{3} & \frac{2}{3} & 1 & \frac{2}{3} & \frac{1}{3} & 0 & 0 \\ 0 & 0 & 0 & \frac{1}{3} & \frac{2}{3} & 1 & \frac{2}{3} & \frac{1}{3} \end{bmatrix} \cdot \begin{bmatrix} 1 \\ 4 \\ 9 \\ 16 \\ 25 \\ 36 \\ 49 \\ 64 \end{bmatrix} = \begin{bmatrix} 31 \\ 52 \\ 112 \end{bmatrix}.$$

The preimage problem given by the vector we ended up with (this ensures that $b \in B$) then has the form

$$b = DF[u] = \begin{bmatrix} 31 \\ 52 \\ 112 \end{bmatrix} = \begin{bmatrix} \frac{1}{3} & \frac{2}{3} & 1 & \frac{2}{3} & \frac{1}{3} & 0 & 0 & 0 \\ 0 & \frac{1}{3} & \frac{2}{3} & 1 & \frac{2}{3} & \frac{1}{3} & 0 & 0 \\ 0 & 0 & 0 & \frac{1}{3} & \frac{2}{3} & 1 & \frac{2}{3} & \frac{1}{3} \end{bmatrix} \cdot v = Wv.$$

We will use SVD of W to find one of the right inverse matrices that constitute a solution. For one particular SVD decomposition, we obtain an explicit solution

$$v = ZS^+P^\top b \doteq \begin{bmatrix} 10.563025 \\ 8.92437 \\ 7.285714 \\ 6.403361 \\ 29.92437 \\ 53.445378 \\ 43.764706 \\ 21.882353 \end{bmatrix}.$$

To verify it, we compute
$$Wv \doteq \begin{bmatrix} 31 & 52 & 112 \end{bmatrix}^\top,$$
and comparing it to b we see that we found a solution to the preimage problem.

In Example A1, we obtained the same particular solution while using the explicit formula from Theorem 1 as well as while using the right inverse given by SVD of W (Corollary 1) and the right inverse given by Lemma 8. As SVD is widely used to compute pseudoinverses, this was not too surprising and our software is no exception.

Examples A2 and A3 relate to Section 4 and show that under certain conditions, the inverse F-transform is one of the possible solutions to the preimage problem. The former example illustrates the degenerate case where not all points of the universe are covered by basic functions (are not connected with at least one node), the latter example uses non-convex basic functions.

Example A2. Let $Y = \{3, 4, 6\}$ be a selected subset of nodes of the set $X = \{1, 2, 3, 4, 5, 6, 7, 8\}$,
$$\forall t \in Y \colon A_t(x) = \begin{cases} 1 & t = x \\ 0 & t \neq x \end{cases}$$
be singleton basic functions on X associated with elements of Y that determine the closeness matrix
$$W = \begin{bmatrix} 0 & 0 & 1 & 0 & 0 & 0 & 0 & 0 \\ 0 & 0 & 0 & 1 & 0 & 0 & 0 & 0 \\ 0 & 0 & 0 & 0 & 0 & 1 & 0 & 0 \end{bmatrix},$$
and the scaling matrix
$$D = \begin{bmatrix} 1 & 0 & 0 \\ 0 & 1 & 0 \\ 0 & 0 & 1 \end{bmatrix}.$$

Let u be a particular function on X with a vector form
$$u = \begin{bmatrix} 0 & 0 & 9 & 16 & 0 & 36 & 0 & 0 \end{bmatrix}^\top.$$

The components of the direct F-transform of u corresponding to all A_t's ($t \in Y$) form the vector:
$$F[u] = \begin{bmatrix} 9 & 16 & 36 \end{bmatrix}^\top.$$

Obviously, it holds that $DF[u] = Wu$, i.e.,
$$\begin{bmatrix} 1 & 0 & 0 \\ 0 & 1 & 0 \\ 0 & 0 & 1 \end{bmatrix} \cdot \begin{bmatrix} 9 \\ 16 \\ 36 \end{bmatrix} = \begin{bmatrix} 0 & 0 & 1 & 0 & 0 & 0 & 0 & 0 \\ 0 & 0 & 0 & 1 & 0 & 0 & 0 & 0 \\ 0 & 0 & 0 & 0 & 0 & 1 & 0 & 0 \end{bmatrix} \cdot \begin{bmatrix} 0 \\ 0 \\ 9 \\ 16 \\ 0 \\ 36 \\ 0 \\ 0 \end{bmatrix} = \begin{bmatrix} 9 \\ 16 \\ 36 \end{bmatrix}.$$

The preimage problem given by the vector we ended up with then has the form
$$b = DF[u] = \begin{bmatrix} 9 \\ 16 \\ 36 \end{bmatrix} = \begin{bmatrix} 0 & 0 & 1 & 0 & 0 & 0 & 0 & 0 \\ 0 & 0 & 0 & 1 & 0 & 0 & 0 & 0 \\ 0 & 0 & 0 & 0 & 0 & 1 & 0 & 0 \end{bmatrix} \cdot v = Wv.$$

The inverse F-transform of u with respect to W has the form

$$\hat{F}[u] = W^\top F[u] = \begin{bmatrix} 0 & 0 & 9 & 16 & 0 & 36 & 0 & 0 \end{bmatrix}^\top.$$

Since this vector is equal to u, we obtain that $W\hat{F}[u] = Wu$, so $\hat{F}[u]$ is a solution to the preimage problem. Note that W is such that $WW^\top = I = D$.

Example A3. Let $Y = \{3, 4, 6\}$ be a selected subset of nodes of the set $X = \{1, 2, 3, 4, 5, 6, 7, 8\}$ and let the closeness be determined by the matrix

$$W = \begin{bmatrix} 1 & 1 & 1 & 0 & 0 & 0 & 0 & 0 \\ 0 & 0 & 0 & 1 & 1 & 0 & 0 & 0 \\ 0 & 0 & 0 & 0 & 0 & 1 & 1 & 1 \end{bmatrix},$$

and the corresponding scaling matrix

$$D = \begin{bmatrix} 3 & 0 & 0 \\ 0 & 2 & 0 \\ 0 & 0 & 3 \end{bmatrix}.$$

Let u be a particular function on X with a vector form

$$u = \begin{bmatrix} 0 & 0 & 9 & 16 & 0 & 36 & 0 & 0 \end{bmatrix}^\top.$$

The components of the direct F-transform of u corresponding to all A_t's ($t \in Y$) form the vector:

$$F[u] = \begin{bmatrix} 3 & 8 & 12 \end{bmatrix}^\top.$$

Obviously, it holds that $DF[u] = Wu$, i.e.,

$$\begin{bmatrix} 3 & 0 & 0 \\ 0 & 2 & 0 \\ 0 & 0 & 3 \end{bmatrix} \cdot \begin{bmatrix} 3 \\ 8 \\ 12 \end{bmatrix} = \begin{bmatrix} 1 & 1 & 1 & 0 & 0 & 0 & 0 & 0 \\ 0 & 0 & 0 & 1 & 1 & 0 & 0 & 0 \\ 0 & 0 & 0 & 0 & 0 & 1 & 1 & 1 \end{bmatrix} \cdot \begin{bmatrix} 0 \\ 0 \\ 9 \\ 16 \\ 0 \\ 36 \\ 0 \\ 0 \end{bmatrix} = \begin{bmatrix} 9 \\ 16 \\ 36 \end{bmatrix}.$$

The preimage problem given by the vector we ended up with then has the form

$$b = DF[u] = \begin{bmatrix} 9 \\ 16 \\ 36 \end{bmatrix} = \begin{bmatrix} 1 & 1 & 1 & 0 & 0 & 0 & 0 & 0 \\ 0 & 0 & 0 & 1 & 1 & 0 & 0 & 0 \\ 0 & 0 & 0 & 0 & 0 & 1 & 1 & 1 \end{bmatrix} \cdot v = Wv.$$

The inverse F-transform of u with respect to W has the form

$$\hat{F}[u] = W^\top F[u] = \begin{bmatrix} 3 & 3 & 3 & 8 & 8 & 12 & 12 & 12 \end{bmatrix}^\top.$$

We see that

$$W\hat{F}[u] = \begin{bmatrix} 1 & 1 & 1 & 0 & 0 & 0 & 0 & 0 \\ 0 & 0 & 0 & 1 & 1 & 0 & 0 & 0 \\ 0 & 0 & 0 & 0 & 0 & 1 & 1 & 1 \end{bmatrix} \cdot \begin{bmatrix} 3 \\ 3 \\ 3 \\ 8 \\ 8 \\ 12 \\ 12 \\ 12 \end{bmatrix} = \begin{bmatrix} 9 \\ 16 \\ 36 \end{bmatrix} = Wu,$$

so $\hat{F}[u]$ is a solution to the preimage problem. Note that again W is such that $WW^\top = D$.

Examples A4 and A5 relate to Section 5 and illustrate that creating a compatible set of basic functions leads to a solution to the preimage problem. The latter example shows a drawback of requiring reflexivity of closeness.

Example A4. *Let* $X = \{1, 2, 3, 4, 5, 6, 7, 8, 9\}$ *be the universe with a selected subset* $Y = \{2, 5, 8\}$ *of nodes and let the closeness be determined by the matrix*

$$W = \begin{bmatrix} \frac{1}{2} & \frac{3}{4} & \frac{1}{2} & 0 & 0 & 0 & 0 & 0 & 0 \\ 0 & 0 & 0 & \frac{1}{2} & \frac{3}{4} & \frac{1}{2} & 0 & 0 & 0 \\ 0 & 0 & 0 & 0 & 0 & 0 & \frac{1}{2} & \frac{3}{4} & \frac{1}{2} \end{bmatrix},$$

and the corresponding scaling matrix

$$D = \begin{bmatrix} \frac{7}{4} & 0 & 0 \\ 0 & \frac{7}{4} & 0 \\ 0 & 0 & \frac{7}{4} \end{bmatrix}.$$

Let $u = x^2$ *be a particular function on X which has a vector form*

$$u = \begin{bmatrix} 1 & 4 & 9 & 16 & 25 & 36 & 49 & 64 & 81 \end{bmatrix}^\top.$$

The components of the direct F-transform of the function u corresponding to all A_t*'s* ($t \in Y$) *form the vector:*

$$F[u] = \begin{bmatrix} \frac{32}{7} & \frac{179}{7} & \frac{452}{7} \end{bmatrix}^\top.$$

Obviously, it holds that $DF[u] = Wu$, *i.e.,*

$$\begin{bmatrix} \frac{7}{4} & 0 & 0 \\ 0 & \frac{7}{4} & 0 \\ 0 & 0 & \frac{7}{4} \end{bmatrix} \cdot \begin{bmatrix} \frac{32}{7} \\ \frac{179}{7} \\ \frac{452}{7} \end{bmatrix} = \begin{bmatrix} \frac{1}{2} & \frac{3}{4} & \frac{1}{2} & 0 & 0 & 0 & 0 & 0 & 0 \\ 0 & 0 & 0 & \frac{1}{2} & \frac{3}{4} & \frac{1}{2} & 0 & 0 & 0 \\ 0 & 0 & 0 & 0 & 0 & 0 & \frac{1}{2} & \frac{3}{4} & \frac{1}{2} \end{bmatrix} \cdot \begin{bmatrix} 1 \\ 4 \\ 9 \\ 16 \\ 25 \\ 36 \\ 49 \\ 64 \\ 81 \end{bmatrix} = \begin{bmatrix} 8 \\ \frac{179}{4} \\ 113 \end{bmatrix}.$$

The preimage problem given by the vector we ended up with then has the form

$$b = \begin{bmatrix} 8 \\ \frac{179}{4} \\ 113 \end{bmatrix} = \begin{bmatrix} \frac{1}{2} & \frac{3}{4} & \frac{1}{2} & 0 & 0 & 0 & 0 & 0 & 0 \\ 0 & 0 & 0 & \frac{1}{2} & \frac{3}{4} & \frac{1}{2} & 0 & 0 & 0 \\ 0 & 0 & 0 & 0 & 0 & 0 & \frac{1}{2} & \frac{3}{4} & \frac{1}{2} \end{bmatrix} \cdot v = Wv.$$

Let us consider, e.g., the following matrix M satisfying $WM^\top = I$:

$$M = \begin{bmatrix} \frac{1}{2} & \frac{2}{3} & \frac{1}{2} & 0 & 0 & 0 & 0 & 0 & 0 \\ 0 & 0 & 0 & \frac{1}{2} & \frac{2}{3} & \frac{1}{2} & 0 & 0 & 0 \\ 0 & 0 & 0 & 0 & 0 & 0 & \frac{1}{2} & \frac{2}{3} & \frac{1}{2} \end{bmatrix}.$$

Then the vector

$$v = M^\top b = \begin{bmatrix} \frac{1}{2} & \frac{2}{3} & \frac{1}{2} & 0 & 0 & 0 & 0 & 0 \\ 0 & 0 & 0 & \frac{1}{2} & \frac{2}{3} & \frac{1}{2} & 0 & 0 \\ 0 & 0 & 0 & 0 & 0 & \frac{1}{2} & \frac{2}{3} & \frac{1}{2} \end{bmatrix}^\top \cdot \begin{bmatrix} 8 \\ \frac{179}{4} \\ 113 \end{bmatrix} = \begin{bmatrix} 4 \\ \frac{16}{3} \\ 4 \\ \frac{179}{8} \\ \frac{179}{6} \\ \frac{179}{8} \\ \frac{113}{2} \\ \frac{226}{3} \\ \frac{113}{2} \end{bmatrix},$$

is a solution to the preimage problem that can be verified by

$$Wv = \begin{bmatrix} \frac{1}{2} & \frac{3}{4} & \frac{1}{2} & 0 & 0 & 0 & 0 & 0 \\ 0 & 0 & 0 & \frac{1}{2} & \frac{3}{4} & \frac{1}{2} & 0 & 0 \\ 0 & 0 & 0 & 0 & 0 & \frac{1}{2} & \frac{3}{4} & \frac{1}{2} \end{bmatrix} \cdot \begin{bmatrix} 4 \\ \frac{16}{3} \\ 4 \\ \frac{179}{8} \\ \frac{179}{6} \\ \frac{179}{8} \\ \frac{113}{2} \\ \frac{226}{3} \\ \frac{113}{2} \end{bmatrix} = \begin{bmatrix} 8 \\ \frac{179}{4} \\ 113 \end{bmatrix} = b.$$

Moreover, it demonstrates the fact that the sets of basic functions A_t's (given by the rows of W) and B_t's (given by the rows of M), $t \in Y$, are compatible.

If we assumed that for each $t \in Y$ it holds $A_t(t) = 1$ (reflexive closeness), the compatible set of convex basic functions satisfying that also for each $t \in Y$ it holds $B_t(t) = 1$, would be uniquely formed by singletons (degenerated case). This is illustrated by Example A5.

Example A5. *Let $X = \{1,2,3,4,5,6,7,8,9\}$ be the universe with the selected subset $Y = \{2,5,8\}$ of nodes and let the closeness be determined by the matrix*

$$W = \begin{bmatrix} \frac{1}{2} & 1 & \frac{1}{2} & 0 & 0 & 0 & 0 & 0 & 0 \\ 0 & 0 & 0 & \frac{1}{2} & 1 & \frac{1}{2} & 0 & 0 & 0 \\ 0 & 0 & 0 & 0 & 0 & \frac{1}{2} & 1 & \frac{1}{2} \end{bmatrix},$$

and the corresponding scaling matrix

$$D = \begin{bmatrix} 2 & 0 & 0 \\ 0 & 2 & 0 \\ 0 & 0 & 2 \end{bmatrix}.$$

Let $u = x^2$ be a particular function on X which has a vector form

$$u = \begin{bmatrix} 1 & 4 & 9 & 16 & 25 & 36 & 49 & 64 & 81 \end{bmatrix}^\top.$$

The components of the direct F-transform of the function u corresponding to all A_ts ($t \in Y$) form the vector:

$$F[u] = \begin{bmatrix} \frac{9}{2} & \frac{51}{2} & \frac{129}{2} \end{bmatrix}^\top.$$

Obviously, it holds that $DF[u] = Wu$, i.e.,

$$\begin{bmatrix} 2 & 0 & 0 \\ 0 & 2 & 0 \\ 0 & 0 & 2 \end{bmatrix} \cdot \begin{bmatrix} \frac{9}{2} \\ \frac{51}{2} \\ \frac{129}{2} \end{bmatrix} = \begin{bmatrix} \frac{1}{2} & 1 & \frac{1}{2} & 0 & 0 & 0 & 0 & 0 & 0 \\ 0 & 0 & 0 & \frac{1}{2} & 1 & \frac{1}{2} & 0 & 0 & 0 \\ 0 & 0 & 0 & 0 & 0 & 0 & \frac{1}{2} & 1 & \frac{1}{2} \end{bmatrix} \cdot \begin{bmatrix} 1 \\ 4 \\ 9 \\ 16 \\ 25 \\ 36 \\ 49 \\ 64 \\ 81 \end{bmatrix} = \begin{bmatrix} 9 \\ 51 \\ 129 \end{bmatrix}.$$

The preimage problem given by the vector we ended up with then has the form

$$b = \begin{bmatrix} 9 \\ 51 \\ 129 \end{bmatrix} = \begin{bmatrix} \frac{1}{2} & 1 & \frac{1}{2} & 0 & 0 & 0 & 0 & 0 & 0 \\ 0 & 0 & 0 & \frac{1}{2} & 1 & \frac{1}{2} & 0 & 0 & 0 \\ 0 & 0 & 0 & 0 & 0 & 0 & \frac{1}{2} & 1 & \frac{1}{2} \end{bmatrix} \cdot v = Wv.$$

Let us consider the following matrix M satisfying $WM^\top = I$ (if we want each of its rows to form a convex fuzzy set with $M^t[t] = 1$, then this is the only option):

$$M = \begin{bmatrix} 0 & 1 & 0 & 0 & 0 & 0 & 0 & 0 & 0 \\ 0 & 0 & 0 & 0 & 1 & 0 & 0 & 0 & 0 \\ 0 & 0 & 0 & 0 & 0 & 0 & 0 & 1 & 0 \end{bmatrix}.$$

Note that if for any $s \in Y \setminus \{t\}$ it holds that $A_t(s) > 0$, then no compatible matrix M exists. The vector

$$v = M^\top b = \begin{bmatrix} 0 & 1 & 0 & 0 & 0 & 0 & 0 & 0 & 0 \\ 0 & 0 & 0 & 0 & 1 & 0 & 0 & 0 & 0 \\ 0 & 0 & 0 & 0 & 0 & 0 & 0 & 1 & 0 \end{bmatrix}^\top \cdot \begin{bmatrix} 9 \\ 51 \\ 129 \end{bmatrix} = \begin{bmatrix} 0 \\ 9 \\ 0 \\ 0 \\ 51 \\ 0 \\ 0 \\ 129 \\ 0 \end{bmatrix},$$

is a solution to the preimage problem that can be verified by

$$Wv = \begin{bmatrix} \frac{1}{2} & 1 & \frac{1}{2} & 0 & 0 & 0 & 0 & 0 & 0 \\ 0 & 0 & 0 & \frac{1}{2} & 1 & \frac{1}{2} & 0 & 0 & 0 \\ 0 & 0 & 0 & 0 & 0 & 0 & \frac{1}{2} & 1 & \frac{1}{2} \end{bmatrix} \cdot \begin{bmatrix} 0 \\ 9 \\ 0 \\ 0 \\ 51 \\ 0 \\ 0 \\ 129 \\ 0 \end{bmatrix} = \begin{bmatrix} 9 \\ 51 \\ 129 \end{bmatrix} = b.$$

This again demonstrates the fact that the sets of basic functions A_t's (given by the rows of W) and B_t's (given by the rows of M), $t \in Y$, are compatible. More importantly, it demonstrates that requiring $A_t(t) = 1$ and $B_t(t) = 1$ for all node indices $t \in Y$ can lead to an empty set of compatible convex basic functions.

References

1. Janeček, J.; Perfilieva, I. Three Methods of Data Analysis in a Space with Closeness. In *Developments of Artificial Intelligence Technologies in Computation and Robotics, Proceedings of the 14th International FLINS Conference (FLINS 2020), Cologne, Germany, 18–21 August 2020*; World Scientific: Singapore, 2020; pp. 947–955.
2. Belkin, M.; Niyogi, P. Laplacian Eigenmaps for Dimensionality Reduction and Data Representation. *Neural Comput.* **2003**, *15*, 1373–1396. [CrossRef]
3. Janeček, J.; Perfilieva, I. F-transform and Dimensionality Reduction: Common and Different. In *International Summer School on Aggregation Operators 2019*; Springer: Berlin/Heidelberg, Germany, 2019; pp. 267–278.
4. Perfilieva, I. Fuzzy Transforms: Theory and Applications. *Fuzzy Sets Syst.* **2006**, *157*, 993–1023. [CrossRef]
5. St. Luce, S.; Sayama, H. Analysis and Visualization of High-Dimensional Dynamical Systems' Phase Space Using a Network-Based Approach. *Complexity* **2022**, *2022*, 3937475. [CrossRef]
6. Schneider, N.; Wurm, G.; Teiser, J.; Klahr, H.; Carpenter, V. Dense Particle Clouds in Laboratory Experiments in Context of Drafting and Streaming Instability. *Astrophys. J.* **2019**, *872*, 3. [CrossRef]
7. Wang, J.; Wang, L.; Liu, X.; Ren, Y.; Yuan, Y. Color-Based Image Retrieval Using Proximity Space Theory. *Algorithms* **2018**, *11*, 115. [CrossRef]
8. Riesz, F. Stetigkeitsbegriff und abstrakte Mengenlehre. In Proceedings of the International Congress of Mathematicians, Rome, Italy, 6–11 April 1908; Volume 2, pp. 18–24.
9. Perfilieva, I.; Daňková, M.; Bede, B. Towards a Higher Degree F-Transform. *Fuzzy Sets Syst.* **2011**, *180*, 3–19. [CrossRef]
10. Stefanini, L. Fuzzy Transform with Parametric LU-Fuzzy Partitions. In *Computational Intelligence in Decision and Control*; World Scientific: Singapore, 2008; pp. 399–404.
11. Perfilieva, I.; Hurtik, P. The F-Transform Preprocessing for JPEG Strong Compression of High-Resolution Images. *Inf. Sci.* **2021**, *550*, 221–238. [CrossRef]
12. Hurtik, P.; Molek, V.; Perfilieva, I. Novel Dimensionality Reduction Approach for Unsupervised Learning on Small Datasets. *Pattern Recognit.* **2020**, *103*, 107291. [CrossRef]
13. Zakeri, K.A.; Ziari, S.; Araghi, M.A.F.; Perfilieva, I. Efficient Numerical Solution to a Bivariate Nonlinear Fuzzy Fredholm Integral Equation. *IEEE Trans. Fuzzy Syst.* **2019**, *29*, 442–454. [CrossRef]
14. Honeine, P.; Richard, C. Preimage Problem in Kernel-based Machine Learning. *IEEE Signal Process. Mag.* **2011**, *28*, 77–88. [CrossRef]
15. Kwok, J.Y.; Tsang, I.H. The Pre-Image Problem in Kernel Methods. *IEEE Trans. Neural Netw.* **2004**, *15*, 1517–1525. [CrossRef] [PubMed]
16. Janeček, J.; Perfilieva, I. Noise Reduction as an Inverse Problem in F-Transform Modelling. In Proceedings of the International Conference on Information Processing and Management of Uncertainty in Knowledge-Based Systems 2022, Milan, Italy, 11–15 July 2022; Springer: Berlin/Heidelberg, Germany, 2022; pp. 405–417. [CrossRef]
17. Perfilieva, I.; Holčapek, M.; Kreinovich, V. A New Reconstruction from the F-Transform Components. *Fuzzy Sets Syst.* **2016**, *288*, 3–25. [CrossRef]
18. Bakır, G.H.; Weston, J.; Schölkopf, B. Learning to Find Pre-Images. *Adv. Neural Inf. Process. Syst.* **2004**, *16*, 449–456.
19. Blyth, T.S. *Set Theory and Abstract Algebra*; Longmans Mathematical Texts; Longman Publishing Group: Harlow, UK, 1975.
20. Knauer, U.; Knauer, K. *Algebraic Graph Theory: Morphisms, Monoids and Matrices*; De Gruyter: Berlin, Germany, 2019.
21. Golub, G.H.; Van Loan, C.F. *Matrix Computations*; Johns Hopkins University Press: Baltimore, MD, USA, 1996.

Article

An Evolving Fuzzy Neural Network Based on Or-Type Logic Neurons for Identifying and Extracting Knowledge in Auction Fraud [†]

Paulo Vitor de Campos Souza [1,*], Edwin Lughofer [1], Huoston Rodrigues Batista [2] and Augusto Junio Guimaraes [3]

[1] Institute for Mathematical Methods in Medicine and Data Based Modeling, Johannes Kepler University Linz, 4040 Linz, Austria
[2] School of Informatics, Communication and Media, University of Applied Sciences Upper Austria Hagenberg, 4232 Hagenberg im Mühlkreis, Austria
[3] Specialization of Artificial Intelligence and Machine Learning, Pontifical Catholic University of Minas Gerais, Belo Horizonte 30535-901, Minas Gerais, Brazil
* Correspondence: paulo.de_campos_souza@jku.at
[†] This paper is an extended version of our paper published in 19th World Congress of the International Fuzzy Systems Association and the 12th Conference of the European Society for Fuzzy Logic and Technology, IFSA-EUSFLAT 2021, Bratislava, Slovakia, 19–24 September 2021; pp. 314–321.

Citation: de Campos Souza, P.V.; Lughofer, E.; Rodrigues Batista, H.; Junio Guimaraes, A. An Evolving Fuzzy Neural Network Based on Or-Type Logic Neurons for Identifying and Extracting Knowledge in Auction Fraud. *Mathematics* **2022**, *10*, 3872. https://doi.org/10.3390/math10203872

Academic Editors: Antonin Dvorak and Vilém Novák

Received: 20 September 2022
Accepted: 14 October 2022
Published: 18 October 2022

Publisher's Note: MDPI stays neutral with regard to jurisdictional claims in published maps and institutional affiliations.

Copyright: © 2022 by the authors. Licensee MDPI, Basel, Switzerland. This article is an open access article distributed under the terms and conditions of the Creative Commons Attribution (CC BY) license (https://creativecommons.org/licenses/by/4.0/).

Abstract: The rise in online transactions for purchasing goods and services can benefit the parties involved. However, it also creates uncertainty and the possibility of fraud-related threats. This work aims to explore and extract knowledge of auction fraud by using an innovative evolving fuzzy neural network model based on logic neurons. This model uses a fuzzification technique based on empirical data analysis operators in an evolving way for stream samples. In order to evaluate the applied model, state-of-the-art neuro-fuzzy models were used to compare a public dataset on the topic and, simultaneously, validate the interpretability results based on a common criterion to identify the correct patterns present in the dataset. The fuzzy rules and the interpretability criteria demonstrate the model's ability to extract knowledge. The results of the model proposed in this paper are superior to the other models evaluated (close to 98.50% accuracy) in the test.

Keywords: evolving fuzzy neural network; or-neuron; auction fraud; knowledge extraction

MSC: 68T35

1. Introduction

Intelligent machine-learning-based models have assumed a prominent role in solving complex problems common in modern society. Intelligent methods can dynamically solve problems and optimize industry processes, disease diagnoses, or actions related to Internet transactions [1].

With the advent of the Internet in people's lives, and the popularization of e-commerce, a current problem is online shopping fraud. Some websites promote user interaction based on bids placed on a product or service. This online auction encourages users to buy products through bids that can be made remotely, according to the standards set by the platforms that offer this service. However, purchases involving high-value goods or rare items often suffer from fraud attempts. In general, online auction fraud has specific characteristics, ranging from products of dubious origin, suspicious and repetitive bidding on a particular item, overvaluation of a product through constant bidding, and users registering on the platform to cause trouble [2]. These problems can make the experience of buying a product at an online auction appalling, preventing new users from opting for this form of purchase and discrediting the platform that offers such services [3]. Considering the actuality of the problem, some researchers have built a database of factors that exist in auctions to identify fraud during transactions to prevent such problems from occurring [4]. These studies support significant areas of current research, and researchers worldwide are working to identify corresponding patterns and propose solutions, including approaches that use clustering and machine learning [5].

Machine-learning-based models can act to assist in the detection of fraud in online auctions. In particular, a model that can combine the efficiency of training artificial neural networks and interpreting fuzzy systems can solve this problem and extract knowledge of auction fraud through linguistic rules about the problem [6]. Fuzzy neural networks can process input data in fuzzy sets and, using neural network training, provide answers to problems with a high degree of interpretability, because this model is comprised of a fuzzy inference system that can represent data in a linguistic and interpretable way using fuzzy operations [7].

When accompanied by logical interpretability about the functioning of the models that act in this solution, this problem solving can facilitate humans' understanding and acceptance of results. Fuzzy natural logic (FNL) is a formal theory that aims to provide a mathematical model of natural human reasoning whose typical feature is using natural language. Its central basis is based on the results of linguistics, artificial intelligence, and logical analysis of natural language [8].

This paper proposes a new evolving fuzzy neural network architecture based on an empirical data analysis fuzzification process that applies the concepts of self-organized direction-aware data partitioning [9]. Logic neurons based on t-conorm (or-neurons) are in the model's second layer. The model's training is founded on Extreme Learning Machine [10] in the offline phase, and the evolving phase is established on indicator-based Recursive Weighted Least Squares [11]. The three-layer model has a fuzzy inference system connected to a neural aggregation network represented in its third (and last) layers. This combination of factors allows the third layer of the model to provide the network's output (defuzzification process) with answers about frauds in online auctions and fuzzy rules that express the logical and interpretable relations present in the evaluated dataset. The model also includes sophisticated capabilities for the behavior's interpretability throughout its analyses, demonstrating tendencies important to interpretability and information extraction. In addition, this paper proposes a presentation of knowledge about a dataset of fraud in auctions with IF–THEN rules, presenting the logical relationships between the dimensions of the problems and the possible impacts of discovering fraudulent behavior. The behavior of the weights of each dimension linked to the fuzzy sets generated gives an interpretative dimension to the issues present in identifying fraud.

The motivation for applying evolving fuzzy neural networks in dynamic behavior problem areas (such as auction fraud) is connected to the dynamic ability of these types of approaches to identify new behaviors from data, incorporating new neural structures into their architecture. Auction fraud problems have an extremely dynamic behavior, as the types of fraud are constantly changing (either by creating new techniques or when previous frauds are discovered). This evolution of knowledge on the models' components helps solve complex problems that can dynamically change over time. Stream data assessments require models to obtain assertive and adaptive outputs. Thus, evolving fuzzy neural networks act this way, with the advantage of demonstrating the evolution of knowledge and identifying behaviors of the elements of their fuzzy rules (antecedents and consequents).

This paper's main highlight is connected with an assertive and interpretable approach to the problem of fraud in auctions. Relations on validating interpretability criteria and generating fuzzy rules are presented to the reader. Another evolving fuzzy neural network model has already been used to solve these demands [12]. However, this paper presents the first neuro-fuzzy model to solve problems and extract advantageous interpretations of the dataset.

The paper is organized as follows: In Section 2, the literature supplies details about evolving fuzzy neural networks, interpretability of evolving fuzzy systems, and auction fraud. Section 3 presents the model employed in this paper, highlighting its layers, training method, and interpretability criteria applied in the model's results. Section 4 presents the test features, the state-of-the-art model to reveal auction fraud, and the results acquired with due arguments and interpretation. Finally, Section 6 highlights the paper's conclusions and possible future work.

2. Literature Review
2.1. Artificial Neural Network and Applications

Artificial neural networks (ANN) are mathematical models that can learn through computational means and are based on human networks of biological neurons. These networks process information using storage and processing structures, with each processor element representing a neuron. They are fed a set of weighted inputs, and their output is obtained by applying an activation function to simulate the degree of stimulation of the biological neuron in order to obtain the responses [13]. According to Fausett and others [14], ANN is a set of computational procedures that provide a mathematical representation inspired by the neural composition of intelligent organisms and that gain knowledge through expertise, allowing tasks typically performed by reasonable beings to be performed in computational environments.

Neural networks are constructed of many interconnected processing units known as neurons. This method is based on the nature of the human brain and how learning and problem solving work. Weighted bonds called synapses connect artificial neurons. Each neuron receives multiple input signals and generates an output signal [13].

Since the 1980s, studies on neural networks have undergone significant evolution, and this area of research has become well-known, either because of the promising features introduced by neural network models or because of the general technological requirement of implementation that allows the audacious development of applications of architectures with parallel neural networks in dedicated hardware, thus acquiring excellent fulfillment of these systems (foremost to conventional methods). Deep neural networks represent the state of the art in neural network development (or Deep Learning) [15]. Where the behavior of variables is not purely known, ANNs are frequently employed to solve complicated issues. Their capacity to learn from examples and generalize the knowledge acquired, producing a nonlinear model, is one of their key properties, which greatly increases the effectiveness of their use in spatial analysis [13].

The selection of the optimum architecture is one of the most challenging aspects of using neural networks since it is experimental and takes a long time to complete. The procedure must be used in practice to evaluate multiple learning models as well as the many topology configurations a network might have to address a particular issue [13].

In two different ways, a neural network approximates the human brain: information is obtained through knowledge stages, and synaptic weights are employed to gather experience. The existing connection between neurons is referred to as a synapse. Values called synaptic weights are assigned to the connections. This demonstrates that artificial neural networks are made up of a succession of artificial (or virtual) neurons that are coupled to form a system of processing components. Their acknowledgments are determined by how the model's underlying structures are linked, and their parameters are predicted [13].

At least one hidden layer is required in ANNs. If the model has multiple hidden layers, each one is accountable for increasing its capacity. The ANN architecture is capable of executing activities that aid in producing model outputs. The neurons in the hidden layers can be activated using activation functions (e.g., Gaussian, triangular, sigmoidal, sine, and hyperbolic tangent). These activation functions are in charge of introducing the nonlinear aspect of the model's activities. Backpropagation or Gradient Descent [16], Extreme Learning Machine [10], and others are examples of training methods for updating model parameters. An example of an artificial neural network is shown in Figure 1.

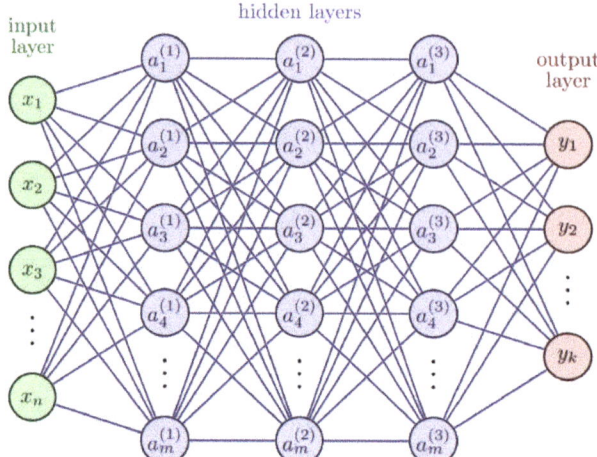

Figure 1. Artificial neural network with five layers.

2.2. Fuzzy Neural Network

Fuzzy neural networks are based on the theory of fuzzy sets and are trained by learning algorithms derived from neural networks. Being able to unify in a model the generalization of parameters and the training capacity provided by artificial neural networks together with the interpretability characteristics of fuzzy systems [17], fuzzy neural networks can work to solve problems of different varieties, making it easier to find solutions for a parameter update in the smart models. By interpreting aspects of the user's database of the problem, they identify existing patterns that make up a set of fuzzy rules capable of transforming numerical information into linguistic contexts that can be interpreted by people who do not necessarily know the main techniques of artificial intelligence. Incorporating fuzzy logic into the neural network can alleviate the insufficiency of each technique, approaching them as more economical, robust, and simple to understand problems.

These models have multilayer architectures, commonly, the three-layer ones stand out, but there may be differences in their architecture. Fuzzy neural networks have also been used for the detection of Pulmonary Tuberculosis [18]. They are also present in the detection of COVID-19 cases from chest X-rays [19]. Most of the models listed above have differences in their architecture. Each of the functions of these layers incorporates the concepts of fuzzification–defuzzification systems and artificial neural networks. The first layer is the one that receives the input data, converting them into fuzzy logic neurons. Fuzzy versions of c-means, adaptive-network-based fuzzy inference systems, and cloud clustering are usually implemented [7]. These techniques usually have at least one hidden layer. In the training algorithms, models based on backpropagation, genetic and evolutionary models, and Extreme Learning Machines stand out [7]. A model with an optimized structure determines its architecture from the fuzzy neural network and an output layer, which is also known as the defuzzification layer.

The architecture of a fuzzy neural network can be expressed in layers, where each one is responsible for a function in the model. Figure 2 represents a fuzzy neural networks in layers where one is responsible for the fuzzification process, the other takes the aggregation of fuzzy neurons, and the last layer is responsible for the model outputs. Recent proposals for fuzzy neural networks have addressed time-series forecasting [20,21], harmonic suppression [22], autonomous vehicles [23], and breast cancer [24].

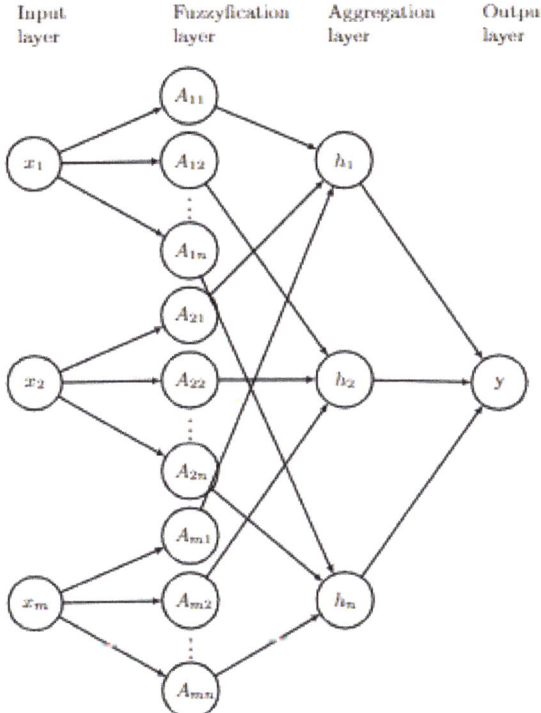

Figure 2. Example of a layered fuzzy neural network.

2.3. Fuzzy Natural Logic

Fuzzy natural logic (FNL) is a class of mathematical theories that includes evaluative expression theory, theory of fuzzy/linguistic IF–THEN rules, and approximate reasoning based on them. The theory of generalized fuzzy quantifiers and their syllogisms, models of natural human reasoning, are also highlighted. FNL aims to develop models of human thought whose most relevant feature is the use of natural language [25].

Many works have been proposed to explore this theory to facilitate the understanding of the results by its users. We highlight works that addressed the theory of intermediate quantifiers in FNL [26], forecasting in analogous time series [27], integration of probabilistic uncertainty to fuzzy systems [28], and interpretable rules with medical data [29], among others.

2.4. Auction Fraud

Online markets present themselves as a very popular way to trade online. E-commerce's main feature is to unite the real and virtual market for the online shopping model, being able to be divided into several sales schemes [30]. The most common business models are Business to Business, which indicates a company that does business with other companies, Business to Consumer, which is the trade carried out directly between the producer, seller, or service provider and the final customer, and Consumer to Consumer, which is based on the direct relationship between consumer and consumer.

According to the Global Payments 2022 Report (https://worldpay.globalpaymentsreport.com/en (accessed on 12 August 2022)) published by FIS Worldpay that examines current and future payment trends in 40 countries across five regions, the global e-commerce market will grow 55% by 2025.

Among the many user benefits, C2C offers minimal costs while maintaining higher margins for sellers and lower prices for buyers. There is also the convenience factor, and customers can list their products online and wait for buyers to come to them. A great example would be an auction.

Large companies promote online auctions, being the only intermediaries to correspond to consumers [30]. The most prominent example is eBay, a successful site since its launch in 1995. Anyone can sign up and start selling or buying. However, fraudsters use it to take advantage by finding alternatives to take advantage of. The leading illicit practices in online auctions are [31]:

- Advertising illegal (stolen) or black market goods;
- Fraud that happens during the bidding period, such as Shill Bidding;
- Post-auction fraud, such as the exaggerated collection of fees, insurance, and even the nondelivery of the purchased goods.

According to the FBI's 2019 Internet Crime Report (FBI's Internet Crime Complaint Center (IC3) in its 2019 Internet Crime Report (available in: https://pdf.ic3.gov/2019_IC3 Report.pdf (accessed on 15 August 2022))), there were 467,361 complaints in 2019—and more than USD 3.5 billion in losses for individual and commercial victims.

The practice of Shill Bidding, the focus of this paper, is when a seller uses a separate account, be it themselves, a friend, or a family member, or someone else, and asks them to bid on their auction to raise the auction price artificially. The item's price is higher than if a legitimate buyer placed a bid and purchased the item. Machine learning techniques have been working on solving fraud identification in auctions. Recent works proposed in [32–35] stand out.

2.5. Evolving Fuzzy Systems and Interpretability

The evolving fuzzy systems are models capable of working with advanced problem solving with a certain degree of interpretability. This problem-interpreting capacity comes from its structure, which can transform data into representations people can read and interpret. A range of models extracts knowledge through fuzzy rules, where each applies techniques and has different architectures for this purpose. The most famous examples of existing evolving fuzzy systems are evolving models based on Takagi–Sugeno, fuzzy classifiers, and fuzzy neural networks.

They differ in the type of fuzzy rules and possible interpretations of them. Several works in the evolving fuzzy system literature work on pattern classification, linear regression, and time-series forecasting (see examples in [36]). In particular, this work deals with interpretable aspects generated by an evolving fuzzy neural network.

Regarding aspects of evolution over time, several proposals were made to facilitate the understanding of the behavior of models as they analyze data. Lughofer [37] proposed criteria that ensure that the fuzzy rules generated can bring certainty to users regarding their actions during the execution of their activities. One should look for more straightforward and distinguishable models, that is, models with a smaller number of fuzzy rules solving the target problems in a coherent and interpretable way, avoiding ambiguities, and that each structure that composes the model avoids redundancy (overlapping of Gaussians may be an example of redundancy present in these models). Other criteria deal with overlapping by evaluating consequents and rule antecedents. The relationship between these two fuzzy rule components determines whether there are inconsistent rules (which can confuse evaluators). Moreover, the evaluation of completeness (a criterion that evaluates the contribution of rules with a significant distance to the sample) and coverage is noteworthy, which verifies whether all samples evaluated by the model are covered by the space of characteristics generated in the fuzzification processes. However, there are also criteria related to assessing the dimensions of the problems. These criteria for interpretability facilitate the identification of dimensions relevant to the issue, which can facilitate understanding a new problem.

On the other hand, this evaluation can also identify less irrelevant features that do not need to be in constructing fuzzy rules as they do not significantly contribute to the resolution of the situation itself. Finally, the criteria for identifying the importance of rules (how much a rule contributes or not to the identification of a target class of the problem), the interpretation of rule consequents (allowing to assess locally and globally the impacts of a fuzzy rule response for the problem solving), and knowledge expansion (identifying when the model identified new patterns and expanded its knowledge base) are fundamental to affirm that a fuzzy neural network model is interpretable. By guaranteeing these criteria, analyzing the generated fuzzy rules with a more outstanding guarantee that they represent solid knowledge about the analyzed problem is possible.

3. Evolving Fuzzy Neural Network Based on Self-Organizing Direction-Aware Data Partitioning (SODA) Approach and Or-Neurons-eFNN-SODA

The development of evolving fuzzy neural networks, models with unprecedented adaptability and freedom, allows acquiring knowledge through the information presented in a data set. This approach assists the precise demonstration of how to build a model able to identify some patterns in a problem analyzed. In this paper, the leading layer of the model proposes a calculation with data density to foster consistent neurons with Gaussian membership function through the idea of or-neurons.

These neurons are responsible for extracting knowledge based on fuzzy rules. The third layer of the model is represented by an artificial neural network model that can process the defuzzification approach. Its training is based on the concept of weight definition through an incremental approach based on recursive least squares. The model's engineering is visible in Figure 3, which is introduced in the following section.

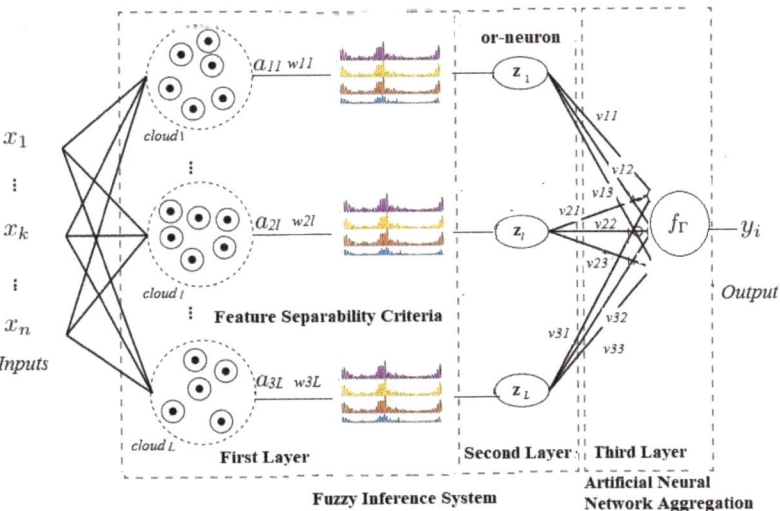

Figure 3. Architecture of the fuzzy neural network proposed in this paper.

3.1. First Layer

The model fuzzification layer is present in the first layer. It is responsible for fuzzifying the data set entries through a clustering process across clouds that identifies them through data density concepts. Therefore, this procedure is responsible for creating Gaussian neurons representing neurons of the first layer. These neurons have weights in the range [0,1] that are defined according to the separability criteria of the dimensions of the problem. For each input variable x_{ij}, L neurons are defined A_{lj}, $l = 1, \ldots L$, whose activation functions are composed of membership functions in the corresponding neurons. The Gaussian neurons created in this layer are expressed by:

$$a_{jl} = \mu_{A_l}, j = 1 \ldots n, l = 1 \ldots L. \tag{1}$$

where a_{jl} (μ_{A_l}) represents the degree of association related to the inputs submitted to the model, N is the number of inputs (features), and L represents the number of neurons. For each input variable x_{ij}, L neurons are defined A_{lj}, $l = 1, \ldots L$, whose activation functions are composed of membership functions in the corresponding neurons [38].

The respective weights of each Gaussian neuron formed by the fuzzification process are created using feature weight separability criteria developed by [39]. These weights are expressed by:

$$w_{il}, i = 1 \ldots N, l = 1 \ldots L \tag{2}$$

The fuzzification process used by the model is described below.

3.1.1. Self-Organizing Direction-Aware Data Partitioning

Self-Organizing Direction-Aware Data Partitioning is a data partitioning approach based on empirical data analysis [40], which can identify peaks/modes of the input frequency and apply them as focal points. Data clouds can be considered a particular variety of clouds that possess an especially peculiar contrast. They are non-parametric, as they do not observe a pre-defined data distribution that is commonly unknown. The technique operates a magnitude component based on a universal distance metric and a directional/angular component based on the cosine similarity [9]. This approach is suitable for the online processing of streaming data.

The definitions of the empirical data analysis [41] operators used in the fuzzification approaches are listed below. SODA, adopted in this paper, acts to update clouds as new information changes its data density. These modifications develop new cloud formats and directly impact the model's formation of the outputs to be acquired. For the SODA approach, consider:

- $\mathbf{x} = \{x_1, x_2, \ldots, x_k\} \in \mathbb{R}^d$: Input variables (where the index k indicates the time instance at which the data point arrives).
- $\mathbf{un} = \{un_1, un_2, \ldots, un_\Psi\} \in \mathbb{R}^d$: Set of unique data point locations.
- $f_1, f_2, \ldots, f_\Psi$: Number of times that different data points occupy the same unique locations.
- K^c: The number of data samples $\{x\}_{K^c}^c$ belonging to the c-th class.
- Un_K^c: The number of unique data samples belonging to the c-th class.
- $\sum_{c=1}^{C} K^c = K$ and $\sum_{c=1}^{C} Un_K^c = Un_K$.

Based on $un_1, un_2, \ldots, un_\Psi$ and $f_1, f_2, \ldots, f_\Psi$, it is possible to reconstruct the dataset $x_1, x_2, \ldots, x_k$ exactly if necessary, regardless of the order of arrival of the data points [41].

The first empirical data analysis operator is cumulative proximity. These elements are defined as the distance between the samples present in the model evaluation. The following is the definition of cumulative proximity [41]:

$$\pi_K(x_i) = \sum_{j=1}^{K} d^2(x_i, x_j); \quad i = 1, 2, \ldots, K \tag{3}$$

where $d(x_i, x_j)$ denotes the distance between x_i and x_j, which can be Euclidean or cosine, among others [42].

The second operator for the determination of data clouds is the unimodal (or local) density, which in turn is determined by [41]:

$$D_K(x_i) = \frac{\sum_{l=1}^{K} \pi_K(x_l)}{2K\pi_K(x_i)} = \frac{\sum_{l=1}^{K} \sum_{j=1}^{K} d^2(x_l, x_j)}{2K \sum_{j=1}^{K} d^2(x_i, x_j)}; \quad i = 1, 2, \ldots, K \tag{4}$$

where, for Euclidean distance, $\sum_{l=1}^{K} d^2(x_i, x_l) = \sum_{l=1}^{K} \|x_i - x_l\|^2$ and $\sum_{l=1}^{K}\sum_{j=1}^{K} d^2(x_l, x_j) = \sum_{l=1}^{K}\sum_{j=1}^{K} \|x_l - x_j\|^2$. It can be simplified using the average of $\{x\}_K$, φ_K and the average scalar product, X_K, as in [43]:

$$\sum_{l=1}^{K} \|x_i - x_l\|^2 = K\left(\|x_i - \varphi_K\|^2 + X_K - \|\varphi_K\|^2\right) \quad (5)$$

$$\sum_{l=1}^{K}\sum_{j=1}^{K} \|x_l - x_j\|^2 = 2K^2\left(X_K - \|\varphi_K\|^2\right) \quad (6)$$

Finally, the third empirical data analysis operator is global density (D_K^G). It is necessary to identify in the data the weighted sum of the local density by a similar occurrence in $f_1, f_2 \ldots, f_{\Psi K}$. It is defined for unique data samples and their corresponding number of repetitions in the data set/stream. It is expressed in (7).

$$D_K^G(un_k) = \frac{f_k}{\sum_{j=1}^{\Psi K} f_j} D_K(un_k) = \frac{f_k}{1 + \frac{\|un_k - \varphi_K\|^2}{X_K - \|\varphi_K\|^2}} \quad (7)$$

The essence of an evolving model is to have its parameters evolve as new samples emerge with new information. The empirical data analysis operators can be updated recursively. This update can be seen below:

$$\varphi_k = \frac{k-1}{k}\varphi_{k-1} + \frac{1}{k}x_k; \quad \varphi_1 = x_1; \quad k = 1, 2, \ldots, K \quad (8)$$

$$X_k = \frac{k-1}{k}X_{k-1} + \frac{1}{k}\|x_k\|^2; \quad X_1 = \|x_1\|^2; \quad k = 1, 2, \ldots, K \quad (9)$$

$$D_K(x_k) = \frac{1}{1 + \frac{\|x_k - \varphi_K\|^2}{X_K - \|\varphi_K\|^2}} \quad (10)$$

The considerable typical distance metric used, Euclidean, was adopted in SODA as the magnitude component and, consequently, can be represented between x_i and x_j by [9]:

$$d_M(x_i) = \|x_i - x_j\| = \sqrt{\sum_{k=1}^{m}(x_{ik} - x_{jk})^2} i, j = 1, 2, \ldots, n \quad (11)$$

The angular component that assumed the ideas of cosine similarity and is provided in the SODA model can be expressed as [9]:

$$d_A(x_i) = \sqrt{1 - \cos(\Theta_{x_i, x_j})} i, j = 1, 2, \ldots, n \quad (12)$$

where $\cos(\Theta_{x_i, x_j}) = \frac{<x_i, x_j>}{\|x_i\|, \|x_j\|}$ expresses the value of the angle between x_i and x_j [9].

Simultaneously, one applies the measurement and angular component values mutually. Considerable problems can be estimated on a 2D plane, called the direction-aware plane [9]. The empirical data analytics operators [40] employed in this approach are the cumulative proximity (3), local density (4), and global density (7).

The recursive update of these parameters also follows the identical concepts as empirical data analysis operators. Their formulation can be seen in Equations (8) and (9). The angular component updates similarly, as expressed below [9]:

$$\varphi_n^A = \frac{n-1}{n}\varphi_{n-1}^A + \frac{1}{n}\frac{x_n}{\|x_n\|}; \quad \varphi_1^A = \frac{x_1}{\|x_1\|} \quad (13)$$

When Euclidean distance is used for Local Density in the SODA approach, it can be represented as follows [9]:

$$\sum_{j=1}^{n} \pi_n^M(x_j) = 2n^2 \left(X_n^M - \|\varphi_n^M\|^2 \right) \tag{14}$$

The initial phases of the SODA algorithm regard, firstly, composing different direction-aware planes of the recognized data samples operating both the magnitude-based and angular-based densities; secondly, distinguishing focal points; and finally, handling the focal points to partition the data space into data clouds [9]. The algorithm is executed in the following steps:

Stage 1—Preparation: Estimate the average values between every pair of data samples, $x_1, x_2, \ldots, x_n$ for both the square Euclidean components, d_M and square angular components, d_A [9].

$$\overline{d}_M^2 = \frac{\sum_{i=1}^{n}\sum_{j=1}^{n} d_M^2(x_i, x_j)}{n^2} = \frac{\sum_{i=1}^{n}\sum_{j=1}^{n} \|x_i - x_j\|^2}{n^2} = 2\left(X_n^M - \|\varphi_n^M\|^2\right) \tag{15}$$

$$\overline{d}_A^2 = \frac{\sum_{i=1}^{n}\sum_{j=1}^{n} d_A^2(x_i, x_j)}{n^2} = \frac{\sum_{i=1}^{n}\sum_{j=1}^{n} \left\|\frac{x_i}{\|x_i\|} - \frac{x_j}{\|x_j\|}\right\|^2}{2n^2} = 1 - \|\varphi_n^A\|^2 \tag{16}$$

After computing the global density, SODA reclassifies the problem samples in a decreasing way and renames them as $\{\widehat{\Phi}_1, \widehat{\Phi}_2, \ldots, \widehat{\Phi}_{n_u n}\}$ [9].

Stage 2—Direction-Aware Plane Projection: The direction aware projection process starts with the unique data sample with the highest global density, namely $\overline{\jmath}_1^-$. It is initially set to be the first reference, $\varphi_1 = \widehat{\Phi}_j$, which is also the origin point of the first direction-aware plane, denoted by P1 ($\Psi = 1$, and Ψ is the number of existing direction-aware planes in the data space. The following rule can describe the second step of the algorithm [9]:

Condition 1 $\quad \begin{array}{c} IF\left(\frac{d_M(\varphi_l, \widehat{\Phi}_j)}{\overline{d}_M} < \frac{1}{\vartheta}\right) AND \left(\frac{d_A(\varphi_l, \widehat{\Phi}_j)}{\overline{d}_A} < \frac{1}{\vartheta}\right) \\ THEN\left(\mathbf{P}_l = \widehat{\Phi}_j\right) \end{array} \tag{17}$

where ϑ assists in the deduction of the granularity of the cloud used. When considerable direction-aware planes satisfy the Equation (17) criteria at the same time, $\widehat{\Phi}_j$ will be assigned to the one nearest to them according to the following equation [9]:

$$i = \arg\min_{l=1,2,\ldots,\Psi} \left(\frac{d_M(\varphi_l, \widehat{\Phi}_j)}{\overline{d}_M} + \frac{d_A(\varphi_l, \widehat{\Phi}_j)}{\overline{d}_A} \right) \tag{18}$$

In this step, the mean, the support (number of samples of the problem), and the sum of the global density (φ_i, Sp_i and D_i, respectively) are updated as follows [9]:

$$\varphi_i = \frac{S_i}{S_i+1}\varphi_i + \frac{1}{S_i+1}\widehat{\Phi}_j \tag{19}$$

$$Sp_i = Sp_i + 1 \tag{20}$$

$$\mathbf{D}_i = \mathbf{D}_i + D_n^G\left(\widehat{\Phi}_j\right) \tag{21}$$

Nonetheless, if Equation (17) is not fulfilled, the parameters involved in the analysis will be updated after the creation of a new direction-aware plane ($P_{\Psi+1}$) and new references, as follows [9]:

$$\Psi = \Psi + 1 \tag{22}$$

$$\varphi_\Psi = \widehat{\Phi}_j \tag{23}$$

$$S_\Psi = 1 \tag{24}$$

$$\mathbf{D}_\Psi = D_n^G\left(\hat{\mathbf{\Phi}}_j\right) \quad (25)$$

This procedure occurs until all problem samples are organized. In this situation, some data samples may be located on several direction-aware planes simultaneously, depending on the behavior of the problem. In this sense, the final establishment of these samples is defined by the distances between them and the points of origin of the following direction-aware plans [9].

Stage 3—Identifying the Focal Points: For each direction-aware plane, denoted as P_β, find the neighboring direction-aware planes ($\{\mathbf{P}\}_\beta^n, l = 1, 2, \ldots, L, l \neq \beta$). The subsequent association can define the rule for determining new plans [9]:

Condition 2
$$IF \left(\frac{d_M(\varphi_\beta, \varphi_l)}{\bar{d}_M} \leq \frac{2}{\bar{\theta}}\right) AND \left(\frac{d_A(\varphi_\beta, \varphi_l)}{\bar{d}_A} \leq \frac{2}{\bar{\theta}}\right) \\ THEN\left(\{\mathbf{P}\}_\beta^n = \{\mathbf{P}\}_\beta^n \cup \{\mathbf{P}_l\}\right) \quad (26)$$

The central mode/peak of data density ($\mathbf{P}_\beta$) will be selected if the following equation is attended (assuming the corresponding $\mathbf{D}$ of $\{\mathbf{P}\}_\beta^n, l = 1, 2, \ldots, \Psi, l \neq \beta$ is expressed by $\{\mathbf{D}\}_\beta^n$ [9]):

Condition 3 $IF\left(\mathbf{D}_\beta > \max\left(\{\mathbf{D}\}_\beta^n\right)\right) THEN\left(\mathbf{P}_\beta \text{ is a mode/peak of } \mathbf{D}\right)$ (27)

All peaks are found when conditions 2 and 3 perform mutually.

Stage 4—Forming Data Clouds: After all the direction-aware planes encountering the modes/peaks of the data density are appointed, it reflects their origin points, represented by φ_o, as the focal points and uses them to form data clouds according to a Voronoi tessellation [44]. The term that represents the data cloud is suggested as follows [9]:

Condition 4
$$IF \left(\diamondsuit = \arg\min_{j=1,2,\ldots,\zeta} \left(\frac{d_M\left(x_l, \varphi_j^o\right)}{\bar{d}_M} + \frac{d_A\left(x_l, \varphi_j^o\right)}{\bar{d}_A}\right)\right) \\ THEN(x_l \text{ is assigned to the } \diamondsuit^{th} \text{ data cloud}) \quad (28)$$

where ζ is seen as the number of focal points. The concept of clouds is equivalent to clusters. Regardless, there are distinctions because they are non-parametric, do not have a typical shape, and can express any real data distributions following local density criteria [45].

This approach can also perform with the evolution of their clouds due to the ability to update their parameters recursively. For the evolving strategy used in SODA, some actions are essential for new samples to influence changes in the data clouds. Therefore, the following steps, which are related to the steps listed previously, help in describing the update of the SODA fuzzification method [9]:

Stage 5—Update Problem Parameters: For each new sample introduced to training (x_{n+1}, φ_n^M, φ_n^A, and X_n^M), they are updated by Equations (9), (13), and (14). Likewise, the Euclidean components and the angular components between the new sample and the established centers of the direction-aware planes are also updated for each new sample employing Equations (11) and (12) and are currently described by $d_M(x_{n+1}, \varphi_l)$ and $d_A(x_{n+1}, \varphi_l)$ for $l = 1, 2, \ldots, \Psi$, sequentially.

At this point in the procedure, the direction-aware plane projection stage is triggered, letting a joint analysis of Condition 1 and Equation (18) determine whether the new sample belongs to an actual data set (model parameters are updated based on Equations (19) and (20), or, if it is a new sample with different behavior, consequently forming the demand to produce a new direction-aware plane (and hence the parameters will be updated based on Equations (22), (23) and (24)) [9].

Stage 6—The Fusion of Overlapping Direction-Aware Planes: After the realization of Stage 5, a new requirement is examined on the fuzzification approach, recognizing strongly overlapping direction-aware planes [9]:

Condition 5
$$IF\left(\frac{d_M(\varphi_i,\varphi_j)}{\bar{d}_M} < \frac{1}{2\theta}\right) AND \left(\frac{d_A(\varphi_i,\varphi_j)}{\bar{d}_A} < \frac{1}{2\theta}\right) \quad (29)$$
$$THEN(\mathbf{P}_i \text{ and } \mathbf{P}_j \text{ are strongly overlapping})$$

This situation is not advisable, so these circumstances are solved by assembling a new direction-aware plane on P_j by combining the analyzed direction-aware plane. This merger criterion is described by [9]:

$$\Psi = \Psi - 1 \quad (30)$$

$$\varphi_j = \frac{Sp_j}{Sp_j + Sp_i}\varphi_j + \frac{S_i}{S_j + S_i}\varphi_i \quad (31)$$

$$Sp_j = Sp_j + Sp_i \quad (32)$$

The procedure occurs until all the direction-aware planes are not highly overlapping. This technique occurs by eliminating the parameters of P_i and returning to Stage 5. This flow occurs until the conditions reported in Condition 5 are not satisfied, thus flowing to the final stage of the evolving fuzzification process [9]. This situation is essential for using the SODA technique, it is a benchmark for certifying the interpretability of the results.

Stage 7—Forming Data Clouds: Once all samples are examined, SODA defines all the focal points of the existent centers in the direction-aware planes resulting from Stage 6. The subsequent steps are correlated to the global density estimation of the centers of the direction-aware plane through Equation (7), operating the supports of each direction-aware plane as the number of repetitions to be performed. Therefore, it obtains the global density, which is called $D_n^G(\varphi_l)$ [9]. Second, each direction-aware plane is examined using Condition 2 to find its neighboring direction-aware plane. In this circumstance, Condition 3 is used in order to confirm whether the $D_n^G(\varphi_e)$ underneath investigation is one of the local maxima of $(D_n^G(\varphi_l))$, so that eventually all the identified maximum $D_n^G(\varphi_l)$ zones and the centers of the corresponding direction-aware planes, together with Condition 4, will act in the establishment of a focal point, and hence in the data clouds [9].

It is meaningful that at the first moment of the model's training, the SODA algorithm's demeanor will be similar to the offline behavior of the algorithm (Stages 1–4). From the subsequent data presented to the model, SODA will act in its evolving state, primarily concentrating on the stages 5–7 of the SODA algorithm [9].

Figure 4 presents an example of the construction of clouds and identification of the centers that guide the construction of the structures resulting from the fuzzification process.

3.1.2. Incremental Feature Weights Learning

This paper addresses a different creation in the definition of weights when compared with various models of evolving fuzzy neural networks that define them randomly. Here, the weights are defined by an online and incremental criterion for evaluating the relevance of the problem features when determining the best separation of the target classes. This proposal increases the reducibility of fuzzy rules, as irrelevant weights can be discarded from a visual and operational evaluation and, at the same time, bring more interpretability to the Gaussian neurons of the first layer of the model. It should be noted that this procedure has already been successfully applied in evolving fuzzy neural networks, as in [11,46].

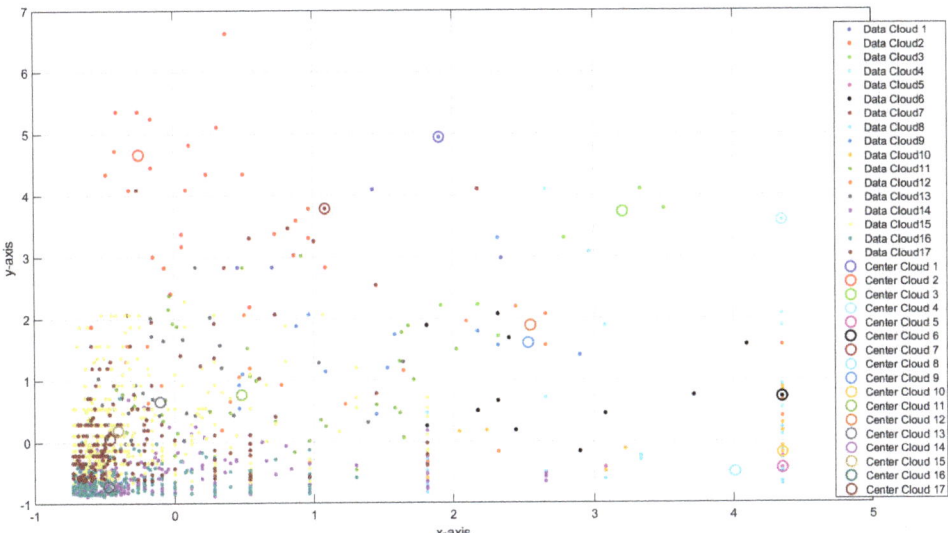

Figure 4. Example of the SODA approach with 17 clouds.

The approach proposed in [39] uses the Dy–Brodley separability criterion [47] for the classes analyzed in the problem. Either compute as a single feature (using each feature (and only one of them)) or discard each feature from the complete set to identify the feature in question that best defines the separation of the classes analyzed. This procedure sets values closer to 1 to features that are more satisfactory in separate classes [11]. On the other hand, dimensions that are ineffective at identifying labels correctly are assigned values close to zero. Accordingly, if a dimension is nonessential to the model, it can be excluded from the model to reduce interpretation complexity.

The class separability criterion can be represented as an extension of Fisher's separability criterion [11]:

$$J = \delta(S_w^{-1} S_b) \tag{33}$$

where $\delta(S_w^{-1} S_b)$ is the sum of the diagonal elements of the matrix $S_w^{-1} S_b$. S_b means the dispersion matrix between classes that estimate the class's dispersion averages to the total average, and S_w denotes the within scattering matrix (that measures the samples' dispersion in their class averages) [11].

This paper uses the leave-one-feature-out approach. It works by discarding each feature from the complete set and then calculates Equation (33) for each obtained subset → again obtaining $J_1, \ldots, J_N$ for N features in the data set. It can be expressed by [39]:

$$w_j = 1 - \frac{J_j - \min_{i=1,\ldots,N} J_i}{\max_{i=1,\ldots,N} J_i} \tag{34}$$

The lower the value of J_i becomes, the more important feature i is, because feature i was discarded. Hence, and seeking comparable importance among all features (relative to the most noteworthy feature, which should acquire a weight of 1), the feature weights are assigned by [39]:

In particular, S_b can be updated by updating class-wise and the overall mean through an incremental mean formula. Recursive covariance update formulas update the covariance matrices per class with rank-1 change for more powerful and faster convergence to the batch calculation [48]. For all formulas and details, check [39].

3.2. Second Layer

The second layer comprises fuzzy logic neurons that use fuzzy operators [49]. The most well-known is the *t-norm*, ($T : [0,1]^2 \rightarrow [0,1]$), which represents the intersection of fuzzy sets, and the *t-conorm* ($S : [0,1]^2 \rightarrow [0.1]$), capable of the union of fuzzy sets. In this paper, these operations are represented by [49]:

$$t(x,y) = x \times y \tag{35}$$

$$s(x,y) = x + y - x \times y \tag{36}$$

Neurons that use t-norm and t-conorm are extremely common in constructing evolving fuzzy neural networks; and-neuron and or-neuron stand out. These fuzzy neural structures, proposed first by Hirota and Pedrycz [50], can aggregate the model inputs (in the case of this paper, the Gaussian neurons of the first layer) with their respective weights. They are considered operational units capable of nonlinear multivariate operations in unit hypercubes ($[0,1] \rightarrow [0,1]^n$) with fuzzy inputs and weights. This process generates a unique value that allows the construction of fuzzy rules in the IF–THEN format. Their interpretability is directly related to the operators applied in the procedures of these fuzzy neuronal structures.

In these neurons with two levels, the first one executes aggregate operations, and the last completes the procedures with fuzzy operators. This operation simplifies the aggregation of the fuzzy input with its respective weight, creating interpretable connectivity between the antecedent connectives [51]. In the case of or-neurons (the neuron structure used in this paper), all or-type connectors are present in the connection of all antecedents of a fuzzy rule generated by this structure [50]. It can be exemplified by:

$$\mathbf{z} = orneuron(\mathbf{w};\mathbf{a}) = S_{i=1}^n (w_i \, t \, a_i) \tag{37}$$

where S is a *t-conorms* and t is *t-norms*, $\mathbf{a} = [a_1, a_2, ..., a_3, ...a_N]$ is the neuron's inputs (values of fuzzy relevance (Gaussian neurons)), and $\mathbf{w} = [w_1, w_2, ..., w_3, ...w_N]$ represents the weights, $\mathbf{a}, \mathbf{w} \in [0,1]^n$, defined in the range between 0 and 1. Finally, $\mathbf{z}$ is the neuron's output, which is also seen as a fuzzy rule. This neuron can generate fuzzy rules as shown below:

$$\begin{aligned}
Rule_1 : \; & If \; x_1 \; is \; A_1^1 \; with \; impact \; w_{11} \ldots \\
& or \; x_2 \; is \; A_1^2 \; with \; impact \; w_{21} \ldots \\
& or \; x_n \; is \; A_1^n \; with \; impact \; w_{n1} \ldots \\
& Then \; y_1 \; is \; v_1 \\
Rule_L : \; & If \; x_1 \; is \; A_L^1 \; with \; impact \; w_{1L} \ldots \\
& or \; x_2 \; is \; A_L^2 \; with \; impact \; w_{2L} \ldots \\
& or \; x_n \; is \; A_L^n \; with \; impact \; w_{nL} \ldots \\
& Then \; y_L \; is \; v_L
\end{aligned} \tag{38}$$

with $A_i^1, \ldots, A_i^n$ fuzzy sets represented as linguistic terms for the n inputs appearing in the ith rules, and $y_1, \ldots, y_L$ are consequent terms (output); L is the number of rules, and $\mathbf{v}$ represents the correspondence value for the output [17].

3.3. Third Layer

The third layer of the fuzzy neural network is composed of the neural network that aggregates all the fuzzy rules generated in the first layer. Its inputs are the fuzzy neurons with their respective weights generated in the training process, to be explained later. An artificial neuron aggregates these fuzzy rules and processes them with weights to generate model outputs. This singleton neural network (composed of a single neuron) has a linear activation function, and a signal function transforms the values obtained by the model into the expected outputs (−1 and 1), indicating whether or not there was fraud in the auction. This defuzzification process can be mathematically described by [38]:

$$y = \Omega \left(\sum_{j=0}^{l} f_\Gamma(z_j, v_j) \right) \quad (39)$$

where $z_0 = 1$, v_0 is the bias, and z_j and $v_j, j = 1, \ldots, l$ are the output of each fuzzy neuron of the second layer and their corresponding weight, respectively. f_Γ represents the linear activation function and Ω the signal function [38].

The linear and signal function are represented, respectively, by:

$$f_\Gamma(\mathbf{z}) = \mathbf{z}^* \mathbf{w} \quad (40)$$

$$\Omega = \begin{cases} 1, & \text{if } \sum_{j=0}^{l} f_\Gamma(z_j, v_j) > 0 \\ -1, & \text{if } \sum_{j=0}^{l} f_\Gamma(z_j, v_j) < 0 \end{cases} \quad (41)$$

3.4. Training

The neural network training process facilitates the definition of existing parameters in an evolving fuzzy neural network. In the model proposed in this article, this training takes place in two stages. The first, offline, is performed using the concept of the Extreme Learning Machine [10], and in the evolving phase of the model, a recursive weighted least squares (RWLS) approach [52] is used. The first process guarantees a definition of weights based on the initial architecture, and the second incrementally updates these weights as new fuzzy rules, and samples with essential information are evaluated. The initial definition of the weights of the third layer carried out in the offline training stage can be expressed by [38]:

$$\vec{v}_k = Z^+ \vec{y}_m \quad \forall m = 1, \ldots, 2 \quad (42)$$

where m is the number of classes, and $Z^+ = Z^T Z$ is the pseudo-inverse of the Moore–Penrose matrix [53] of Z (output of the second layer). This procedure is commonly applied in training artificial neural networks and sets the weights in a single step, allowing the training to be fast and accurate.

In the evolving phase of the model (new incoming online stream samples), the recursive weighted least squares (RWLS) approach is used [52] with formulas for updating the $\vec{v}_k$, and it can be expressed as follow:

$$\eta = \vec{z}^t Q^{t-1} \left(\psi + (\vec{z}^t)^T Q^{t-1} \vec{z}^t \right)^{-1} \quad (43)$$

$$Q^t = (I_{L^t} - \eta^T \vec{z}^t) \psi^{-1} Q^{t-1} \quad (44)$$

$$\vec{v}_k^t = \vec{v}_k^{t-1} + \eta^T (y_m^t - \vec{z}^t \vec{v}_k^{t-1}) \quad (45)$$

where the index k again denotes the class index $k = 1, \ldots, C$. $\vec{z}^t$ denotes the regressor vector of the current sample, η is the current kalman gain vector, $I_{L_s^t}$ is an identity matrix based on the number of neurons in the second layer, and $L_s^t \times L_s^t$; $\psi \in]0,1]$ denotes forgetting factor (1 per default). Q denotes the inverse Hessian matrix $Q = (Z_{sel}^T Z_{sel})^{-1}$ and is set initially as $\omega I_{L_s^t}$, where $\omega = 1000$ [54].

This matrix is directly and incrementally updated by the second equation above without requiring (time-consuming and possibly unpredictable) matrix re-inversion. This procedure identifies existing changes in the dataset and updates the weight values without losing the previous reference. Thus, the concept of memory is applied to the weights, allowing the model's training to be consistent with the premises of evolving fuzzy systems.

The eFNN-SODA can be synthesized, as represented in Algorithm 1. It has one parameter during the initial batch learning phase:

Granularity of the cloud results (grid size), ϑ;

Algorithm 1 eFNN-SODA Training and Update Algorithm

Initial Batch Learning Phase (Input: data matrix X):
(1) Define granularity of the cloud, ϑ.
(2) Extract L clouds in the first layer using the SODA approach (based on ϑ).
(3) Construct L fuzzy neurons with Gaussian membership functions with $\vec{c}$ and σ values derived from SODA.
(4) Calculate the combination (feature) weights $\vec{w}$ for neuron construction using Equation (34).
(5) Construct L logic neurons on the second layer of the network by welding the L fuzzy neurons of the first layer, using or-neurons (Equation (37)) and the centers $\vec{c}$ and widths $\vec{\sigma}$.
(6)
 for $i = 1, \ldots, K$ **do**
  (6.1) Calculate the regression vector $z(x_i)$.
  (6.2) Store it as one row entry into the activation level matrix z.
 end for
(7) Extract activation level matrix z according to the L neurons.
(8) Estimate the weights of the output layer for all classes $k = 1, \ldots, C$ by Equation (42) using z and indicator vectors $\vec{y}_k$.

Update Phase (Input: single data sample $\vec{x}_t$):
(11) Update L clouds and evolving new ones on demand (based on Stages 5, 6, and 7 of the SODA approach).
(12) Update the feature weights $\vec{w}$ by updating the within- and between-class scatter matrix and recalculating Equation (34).
(13) Perform Steps (3) and (5).
(14) Calculate the interpretability criteria (Section 3.5).
(15) Calculate the regression vector $z(\vec{x}_t)$.
(16) Update the output layer weights by Equation (45).

The computational complexity of eFNN-SODA encloses the number of flops demanded to process one single sample through the updated algorithm (second part in Algorithm 1) because this affects the online speed of the algorithm. In this sense, the main steps involved can be listed below:

- SODA algorithm: The complexity of $O(p)$ (p is the dimensionality of the input space when updating with a single sample).
- Or-neurons: Complexity $O(mp)$ (constructing them from m clouds).
- Weights in the first-layer Gaussian neurons: Complexity Kp^3 (where K is the number of classes). That happens because the between- and within-class scatter matrices need to be updated for each class for each feature separately and independently, having a complexity of $O(p^2)$ (matrices have a size of $p \times p$).
- Neuron activation in the third-layer model: The complexity of $O(mp)$ (in each sample, the activation levels to all m neurons (with dimensionality p) require to be estimated).
- Output-layer neuron: The complexity of $O(mp^2)$ (because of the weighted method in each rule individually, demanding this complexity [55].)

3.5. Interpretability Criteria

The interpretability criteria are essential for evaluating the model's behavior throughout the evaluations. In this case, the eFNN-SODA model uses some approaches to validate the interpretability criteria proposed in [37], so that the generated fuzzy rules are reliable. In this case, the model applies some evaluations throughout the experiments to ensure that the generated fuzzy rules can add some knowledge about the evaluated dataset. These procedures are listed below:

Simplicity and Distinguishability: These two criteria verify whether the proposed model is the simplest and has its structures distinguishable during training. This means that the evaluation revolves around low complexity and high accuracy. Regarding the simplicity of the eFNN-SODA, the model is expected to have a more compact structure and a high degree of assertiveness. The criterion defined for the identification of model simplicity (in the comparison between models) can be expressed by:

$$if\{(L_a < L_b) \vee (accuracy_{model_a} \geq accuracy_{model_b})\} \text{ then } model_a \text{ is simpler than } model_b \tag{46}$$

where L_a and L_b are, respectively, the number of fuzzy rules of the $model_a$ and $model_b$. Regarding distinguishability, it is expected to assess whether there is an overlapping in the structures formed in the fuzzification process. The SODA approach uses an assessment of overlapping in the evolution process in its sixth stage, thus ensuring that this situation does not occur. For the evaluation of the distinguishability criterion, eFNN-SODA uses the comparison of similarity between the Gaussian neurons formed in the first layer (termed as $z_i(bef)$ and $z_i(after)$) dimension-wise, and similarity (S_{im}) degree $S_{im}(z_i(bef), z_i(after))$ can be used for calculating an amalgamated value. The degree of change ($\ltimes$) is then presented by [11]:

$$\ltimes (z_i) = 1 - S_{im}(z_i(bef), z_i(after)) \tag{47}$$

$bef = N - n$ and $after = N$, assuming that n new samples have passed the data-stream-based transformation phase with N samples operated so far for model training and adaptation [11].

Therefore, it is feasible to conclude that two rules are only identical if all their antecedent parts are equivalent. The x coordinates of the points of intersection of two Gaussians used as fuzzy sets in the identical antecedent part of the Gaussian rule i (here for the jth) before and after its update can be estimated by [56]:

$$inter_x(1,2) = -\frac{\vec{c}_{bef,j}\vec{\sigma}^2_{after,j} - \vec{c}_{after,j}\vec{\sigma}^2_{bef,j}}{\vec{\sigma}^2_{bef,j} - \vec{\sigma}^2_{after,j}}$$
$$+ -\sqrt{(\frac{\vec{c}_{bef,j}\vec{\sigma}^2_{bef,j} - \vec{c}_{bef,j}\vec{\sigma}^2_{after,j}}{\vec{\sigma}^2_{after,j} - \vec{\sigma}^2_{bef,j}})^2 - \frac{\vec{c}^2_{bef,j}\vec{\sigma}^2_{after,j} - \vec{c}^2_{after,j}\vec{\sigma}^2_{bef,j}}{\vec{\sigma}^2_{after,j} - \sigma^2_{bef,j}}} \tag{48}$$

with $\vec{c}_{bef,j}, \vec{\sigma}_{bef,j}$ being the jth center coordinate and standard deviation of the Gaussian neuron before its update, and $\vec{c}_{after,j}$ and $\vec{\sigma}_{after,j}$ the jth center coordinate and standard deviation of the Gaussian neuron after its update [11].

The maximal membership degree of the two Gaussian membership functions in the intersection coordinates is then used as overlap. Consequently, the similarity degree of the corresponding rules' antecedent parts in the jth dimension is [11]:

$$S_{im}^{bef,after}(j) = \max(\mu_i(inter_x(1)), \mu_i(inter_x(2))) \tag{49}$$

with $\mu_i(inter_x(1))$ being the membership degree of the jth fuzzy set in Rule i in the intersection point $inter_x(1)$. The amalgamation of overall rule antecedent parts leads to the final similarity degree between the rule before and after its update:

$$S_{im}(z_i(bef), z_i(after)) = T_{j=1}^{p} S_{im}^{bef,after}(j) \tag{50}$$

where T denotes a t-norm operator, and p is the number of inputs, as a robust nonoverlap along one single dimension is sufficient for the clouds to not overlap at all [56].

- Consistency, Coverage, and Completeness: The concept of consistency in evolving fuzzy systems is attending when their fuzzy rule-set does not deliver a high noise level or an inconsistently learned output behavior. Therefore, a set of fuzzy rules is considered inconsistent when two or more rules overlap in the antecedents, but not in the consequents. In this paper, the consistency of fuzzy rules (comparing a rule before and after its evolution) can be measured by evaluating the similarity involved in its rule antecedents (S_{ante}) and consequents (S_{cons}). In this case, they can be expressed by [37]:

$$\begin{array}{c} \text{Rule } z_1 \text{ is inconsistent to Rule } z_2 \text{ if and only if} \\ S_{ante}(z_1,z_2) \geqslant S_{cons}(z_1,z_2) \text{ with } S_{ante}(z_1,z_2) \geqslant thr. \end{array} \quad (51)$$

$$\text{where consistency} = \begin{cases} 1 & \text{if Equation (51) is false} \\ 0 & \text{if Equation (51) is true} \end{cases}$$

where S_{ante} and S_{cons} are close to 1 invariably can be assumed to indicate a heightened similarity, and when they are close to 0, a low similarity [37]. thr is a threshold value usually set at 0.8 or 0.9.

The coverage criterion identifies whether there are holes in the resource space by generating undefined input states. This criterion can be solved by applying Gaussian functions, which have unlimited support. In this case, eFNN-SODA guarantees this criterion using this type of function throughout the model's training [37].

Finally, the ϵ-completeness criterion in the eFNN-SODA is defined by [37]:

$$\left(\forall \vec{x} \; \exists i \; \left(z_i = \underset{j=1,\ldots,rl}{T}(\mu_{ij}) > \epsilon \right) \right) \Rightarrow \left(\forall \vec{x} \; \exists i \; (\forall j \; \mu_{ij} > \epsilon) \right) \quad (52)$$

where μ_{ij} is the membership degree of a fuzzy set A_j appearing in the jth antecedent part of the ith rule, rl is the rule length, and $\epsilon = 0.135$ according to definitions made in other research, which is considered an evaluation standard for this criterion [37].

- Antecedent Interpretation: The evaluation of rule antecedents is also a fundamental part of interpreting and validating the results. In the case of eFNN-SODA, this evaluation is performed with the behavior of the evolution of the weights and the evaluation of the similarity of the Gaussian neurons formed in the first layer.
- Consequent Interpretation: The evaluation of rule consequents in eFNN-SODA is performed by graphically evaluating changes in the class to which the rule is connected. This is because the values of $\vec{v}$ can change as the model evaluates new samples, and the consequent of the respective rule can be changed.
- Feature Importance Level: The evaluation of the features of the problem is also a fundamental part of understanding the model's behavior. Graphical analyses of the variations generated by the weights in Equation (34) are obtained, allowing us to identify the behavior of the features throughout the experiment.
- Knowledge Expansion: The expansion of knowledge takes place by evaluating the evolution of fuzzy rules throughout the experiment. The SODA approach acts on the evolution and reduction in irrelevant rules. The eFNN-SODA model can also appreciate this behavior throughout the experiment.

4. Auction Fraud Testing

The following sections presents the details and procedures of the tests that were performed, as well as the models and data set. All tests were performed on a computer with the following settings: Intel (R) Core (TM) i7-6700 CPU 3.40 GHz with 16 GB RAM. In the execution and elaboration of the models present in the tests of this paper, Matlab was used, and for data analysis, the Orange Data Mining tool was used (developed in phyton (https://orangedatamining.com/ (accessed on 17 August 2022))).

4.1. Data Set Features

As mentioned previously, this study aims to analyze Shill Bidding fraud. The analyzed dataset was collected by [4] and published in the Machine Learning Repository—UCI (arch ive.ics.uci.edu/ml/datasets/Shill+Bidding+Dataset (accessed on 17 August 2022)). The data correspond to fraudulent bids on one of the largest auction sites on the Internet, eBay.

The collected database [4], features completed iPhone 7 auctions for three months (March to June 2017). The original dataset contains 12 input features. However, for the studies described in this paper, the dimensions related to personal IDs (Record ID, Auction ID, and Bidder ID) were removed since they are ID values and, therefore, irrelevant to the experiments. The remaining nine dimensions are Bidder Tendency, Bidding Ratio, Successive Outbidding, Last Bidding, Auction Bids, Auction Starting Price, Early Bidding, Winning Ratio, Auction Duration, and Class (1 to normal and −1 to fraud). Figure 5 presents statistical values and the data distribution by class involved in the problem.

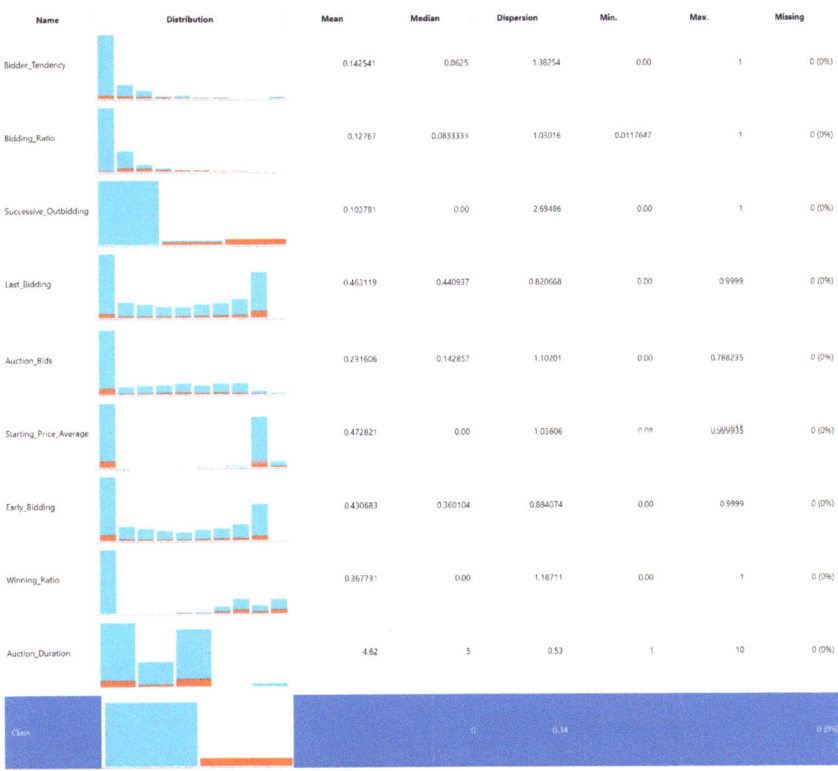

Figure 5. Statistical values and the data distribution—Auction Fraud data set.

As shown in Figure 5, the dataset used in this experiment is imbalanced, making the classification task challenging.

Figure 6 presents a representation of the data using the FreeViz technique [57]. In it, points in the same class attract each other, while those from different classes repel each other, and the resulting forces are exerted on the anchors of the attributes, that is, on unit vectors of each dimensional axis. With this technique, it is possible to identify projections of unit vectors that are very short compared with the others. This indicates that their associated attribute is not very informative for a particular classification task.

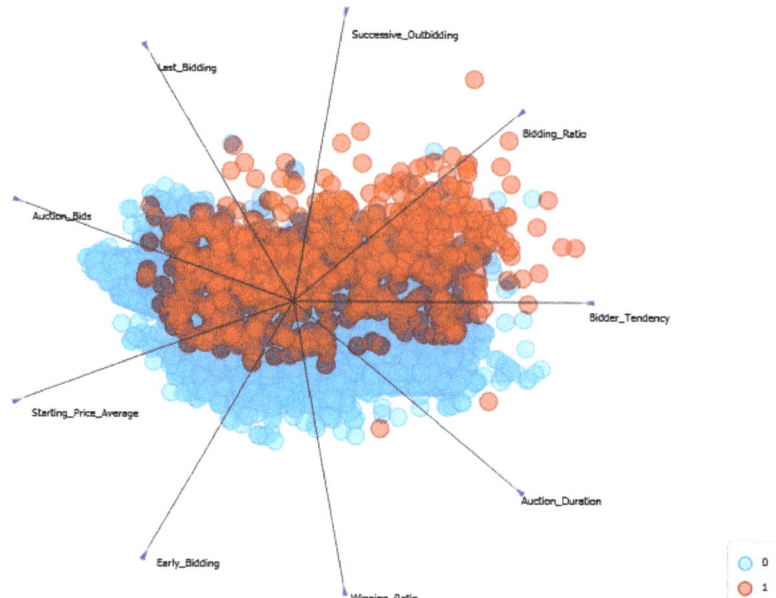

Figure 6. FreeViz projection—Auction Fraud data set.

Based on Figure 6, the most representative dimensions for fraud classification in auctions are successive outbidding, bidding, and winning ratios. In Figure 7, a data visualization according to the Radviz technique [58] is presented to demonstrate a nonlinear multidimensional visualization that can display data defined by three or more variables in a two-dimensional projection. Visualized variables are presented as anchor points evenly spaced around the perimeter of a unit circle. Data instances close to a variable anchor set have higher values for those variables than for others.

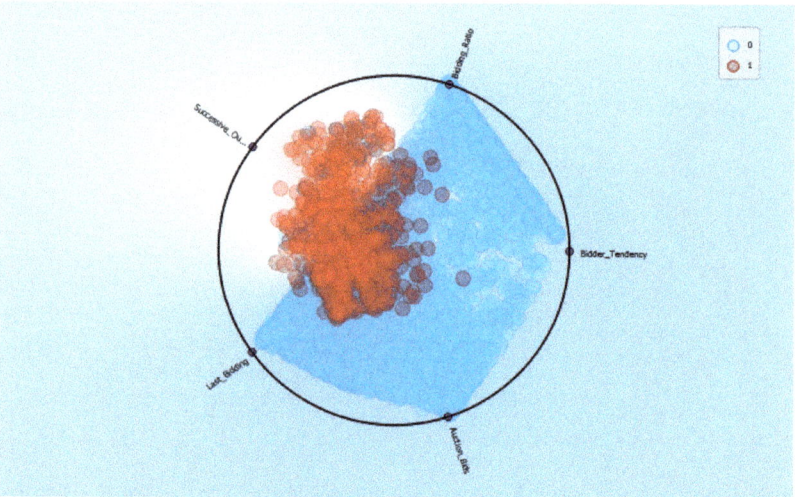

Figure 7. Radviz projection—Auction Fraud data set.

4.2. Models and Hyperparameters

The models used in the experiment are evolving fuzzy systems considered state-of-the-art in classifying binary patterns. They have different architectures and training methods. The parameters of each of these models were defined through pre-tests that used 10-folds in a group of parameters that make up each approach. The models used in the experiment are listed below:

EFDDC—Evolving fuzzy data density cluster. The evolving model uses data-density-based clustering based on empirical data analysis operators and nullneuron. The model's training is based on the Extreme Learning Machine and Bolasso technique to select the best neurons. The model parameters are $\rho = 0.01$, bootstraps $bt = 16$, and consensus threshold $\lambda = 0.7$ (the best parameters are defined between: $\rho = \{0.01, 0.02, 0.03, 0.04\}$, $bt = \{4, 8, 16, 32\}$, and $\lambda = \{0.5, 0.6, 0.7\}$) [59].

EFNHN—Evolving fuzzy neural hydrocarbon network. The model combines an evolving fuzzification technique based on data density (Autonomous Data Partitioning), training based on Extreme Learning Machine, and unineurons. The defuzzification process is based on an Artificial Hydrocarbon network. The model parameter is the Learning rate = 0.1 (the best parameter is defined between: Learning rate = $\{0.01, 0.05, 0.1, 0.2\}$) [60].

EFNNS—Evolving fuzzy neural network and Self-Organized direction aware. The evolving fuzzy neural network model uses the Self-Organized direction aware for fuzzification, unineurons, Extreme Learning Machine, and pruning technique. The only parameter used in the model is the $\vartheta = 3$ (the best parameter is defined between: Learning rate = $\{2, 3, 4, 5\}$) [61].

ALMMo-0*—Autonomous zero-order multiple learning with pre-processing. The model is a neuro-fuzzy approach to autonomous zero-order multiple learning with pre-processing that improves the classifier's accuracy, as it creates stable models. The parameter is radius = $\sqrt{2 - 2\cos(30°)}$ (the best parameter is defined between: radius = $\{\sqrt{2 - 2\cos(15°)}, \sqrt{2 - 2\cos(30°)}, \sqrt{2 - 2\cos(45°)}, \sqrt{2 - 2\cos(60°)}\}$) [62].

ALMMo—Autonomous zero-order multiple learning. A neuro-fuzzy approach for autonomous zero-order multiple models without pre-processing. The parameter is radius = $\sqrt{2 - 2\cos(30°)}$ (the best parameter defined between: radius = $\{\sqrt{2 - 2\cos(15°)}, \sqrt{2 - 2\cos(30°)}, \sqrt{2 - 2\cos(45°)}, \sqrt{2 - 2\cos(60°)}\}$) [63].

eGNN—Evolving Granular Neural Network. A model that uses concepts of hyberboxes and nullnorm. The parameters are Rho = 0.85, eta = 2, hr = 40, Beta = 0, chi = 0.1, zeta = 0.9, c = 0, counter = 1, and alpha = 0 (for this model, the reference values proposed by the author of the model were used. More information can be seen at https://sites.google.com/view/dleite-evolving-ai/algorithms (accessed on 18 August 2022)) [64].

4.3. Evaluation Criteria of Experiments

Evolving fuzzy systems approaches can act in the evaluation of stream data, where each sample is considered separately, and the accuracy of the result is evaluated individually and incrementally added to the final result. In this case, the evaluation of accuracy in trend lines is suitable for this type of context as it allows the evolution of the model results to be seen as new samples are evaluated.

These trend lines were calculated using the following criteria: [65]:

$$Accuracy(K+1) = \frac{Accuracy(K) * K + I_{\hat{y}=y}}{K+1}, \tag{53}$$

where accuracy:

$$Accuracy = \frac{TP + TN}{TP + FN + TN + FP} * 100. \tag{54}$$

where TP = true positive, TN = true negative, FN = false negative, and FP = false positive.

4.4. Results

The results of the evaluation of the trend lines are presented in Figure 8. This evaluation was carried out with 20% of the total samples (1264) for initial training and the rest (5057) for online adaptation, evolution, and evaluation of the model. For the experiments, the models listed above were compared with the results obtained by eFNN-SODA with $\vartheta = 2$ (the values were defined through a cross-validation procedure for $\vartheta = \{2, 3, 4, 5\}$).

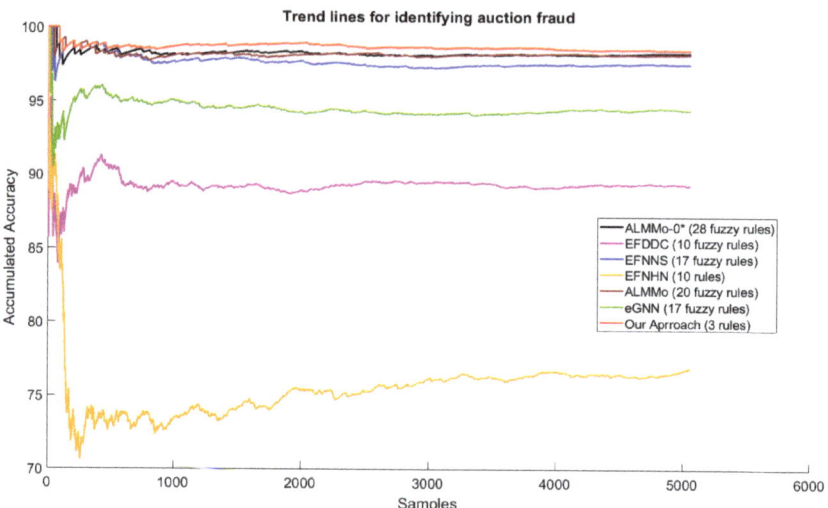

Figure 8. Trend lines for the Auction Fraud dataset.

The results presented in Figure 8 corroborate the efficiency of the approach in correctly classifying the situations involved in the data set that deal with fraud in auctions. The model proposed in this paper could dynamically and more assertively resolve cases of attack or nonexistence in evaluating the samples. Next to the results obtained by eFNN-SODA are the results generated by Autonomous zero-order multiple learning with pre-processing and Autonomous zero-order multiple learning. Numerically, they are similar, with a slight advantage for eFNN-SODA. A little less effective, but with good results, is the evolving fuzzy neural network with Self-Organizing Direction-Aware Data Partitioning model. The other models did not perform as effectively as the eFNN-SODA models and the Autonomous zero-order multiple learning variations.

The eFNN-SODA model solved the problem of fraud in auctions with three rules. In this case, the Self-Organizing Direction-Aware Data Partitioning fuzzification approach did not generate new neurons, as it was only necessary to change centers in the rule antecedents to solve the problem. The three fuzzy rules are presented in the following:

Rule 1. If Bidder Tendency is small with impact 0.41 or Bidding Ratio is small with impact 0.59 or Successive Outbidding is small with impact 1.00 or Last Bidding is medium with impact 0.39 or Auction Bids is small with impact 0.38 or Auction Starting Price is medium with impact 0.38 or Early Bidding is medium with impact 0.38 or Winning Ratio is small with impact 0.53 or Auction Duration is small with impact 0.38 then output is normal.

Rule 2. If Bidder Tendency is medium with impact 0.41 or Bidding Ratio is medium with impact 0.59 or Successive Outbidding is medium with impact 1.00 or Last Bidding is high with impact 0.39 or Auction Bids is high with impact 0.38 or Auction Starting Price is high with impact 0.38 or Early Bidding is high with impact 0.38 or Winning Ratio is medium with impact 0.53 or Auction Duration is medium with impact 0.38 then output is fraud.

Rule 3. If Bidder Tendency is high with impact 0.41 or Bidding Ratio is high with impact 0.59 or Successive Outbidding is high with impact 1.00 or Last Bidding is small with impact 0.39 or Auction Bids is medium with impact 0.38 or Auction Starting Price is small with impact 0.38 or Early Bidding is small with impact 0.38 or Winning Ratio is high with impact 0.53 or Auction Duration is high with impact 0.38 then output is normal.

These rules are seen as the extraction of knowledge from the evaluated data set and can serve as a conceptual analysis of the events involved in identifying or not identifying fraud in auctions. The interpretability criteria help understand and validate the generated knowledge to confirm the efficiency of the generated fuzzy rules.

The first evaluation seeks to verify whether the eFNN-SODA model was the simplest in solving auction fraud problems. This factor is confirmed because, according to Equation (46), the eFNN-SODA model had the lowest number of fuzzy rules and the highest accuracy result compared with the other models used in the experiments. eFNN-SODA assertively solved the target problem with only three fuzzy rules. As a comparison, the different models tested in the experiments generated between 10 and 28 fuzzy rules.

The distinguishability criteria (based on Equation (50)) of the formed solution are presented in Figure 9. In this figure, the evolution of the first layer of Gaussian neurons can be observed, and it is identified that the neuron was the one that suffered the most changes during the experiments. It is responsible for representing when there is no fraud in the auction (the largest class of the data set). Therefore, it is expected that with a more significant number of samples in this context, the changes in clouds will be more remarkable, consequently generating a greater impact on the comparative similarity with their previous version (before the evolution training). In this evaluation, it is possible to identify that the fuzzy rules (presented below) are distinguishable from each other, as they have different antecedents and consequents.

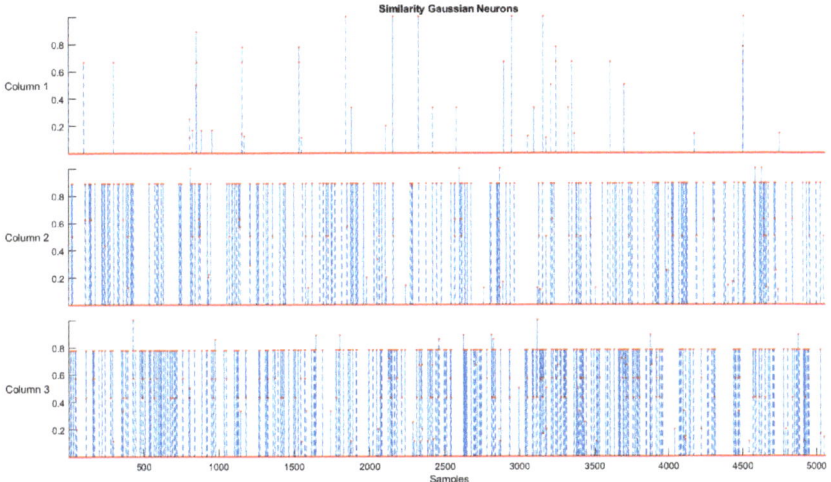

Figure 9. Similarity of Gaussian neurons.

Regarding the overlapping criteria of the generated Gaussians, it can be seen in Figure 10 (generated at the beginning of the training) that the centers of the generated fuzzy neurons are not superimposed. This confirms that there are differences between them. This criterion is also guaranteed by steps 5 and 6 of the Self-Organizing Direction-Aware Data Partitioning fuzzification approach. This figure presents the evaluation concerning the successive outbidding and bidding ratio dimensions (more relevant in the model analysis shown below).

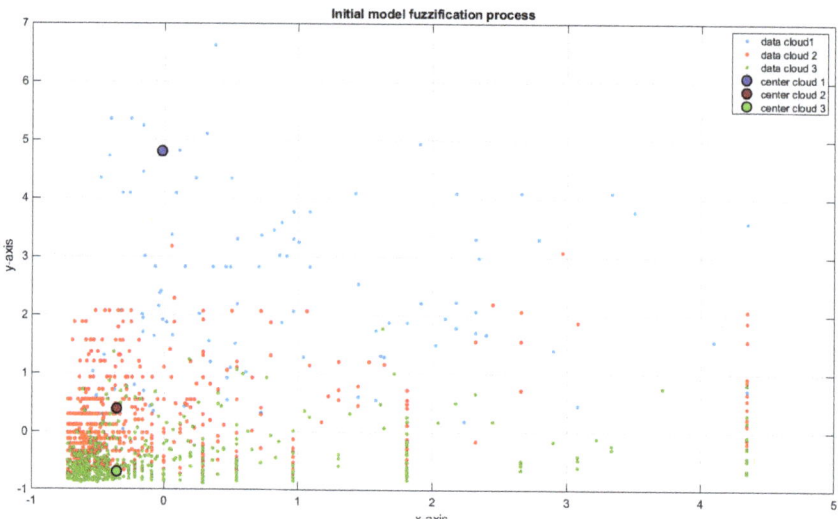

Figure 10. Evaluation of overlapping of the three generated fuzzy neurons during the fuzzification process: successive outbidding versus bidding ratio.

The consistency of the generated fuzzy rules (Equation (51)) is visualized in Figure 11. In this figure, it is possible to see that the consistency relation of the fuzzy rules is violated only in a few evaluations in each of the three fuzzy rules generated. At the end of the experiment, all rules are consistent, concluding that model training corrected these inconsistencies when they appeared.

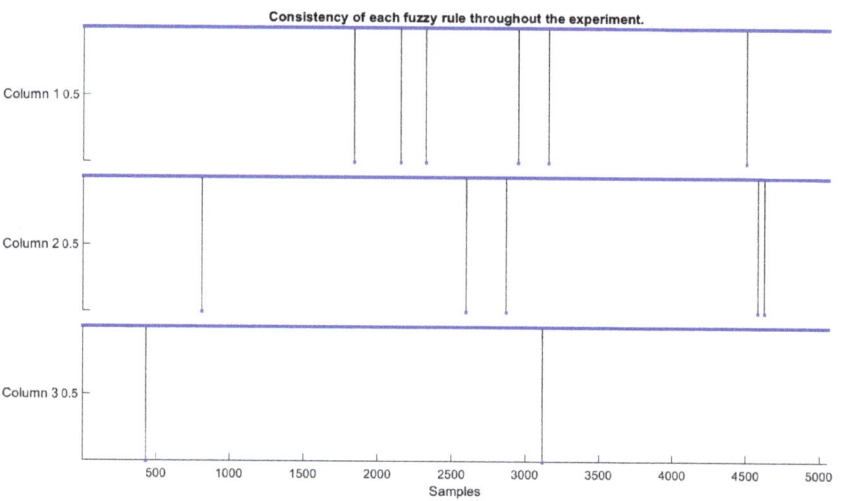

Figure 11. Consistency of fuzzy neurons.

Concerning the criterion of completeness of the fuzzy rules, an evaluation based on $\epsilon - completeness$ was applied during training, according to Equation (52). The results visualized in Figure 12 confirm that the generated fuzzy ruleset meets this validation criterion throughout the experiment.

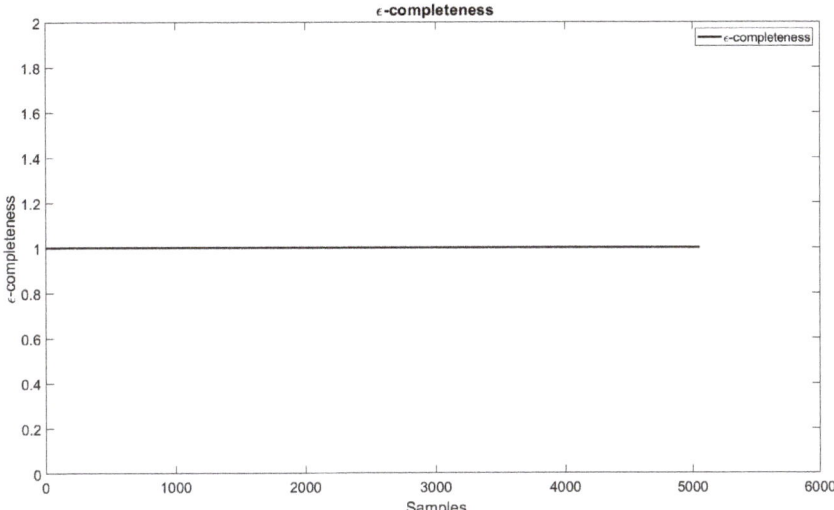

Figure 12. $\epsilon - completeness$ criteria during the training.

Changes in antecedents and consequents of fuzzy rules can also be observed and visualized in eFNN-SODA. Table 1 presents feedback from the model in the last sample evaluation that generated impacts on the first and third fuzzy rules

Table 1. Interpretability concerning (degree of) changes in fuzzy neurons during the evolution phase.

Rule 1 changed with two membership functions, and this impacts changes in the consequent rule by a degree of 45%.
Rule 2 did not change.
Rule 3 changed with two membership functions, and this impacts changes in the consequent rule by a degree of 0.0003%.

The evaluation of the evolving behavior of the problem features is also a relevant part of the interpretability of the results. This variation can be seen in Figure 13 and identifies how each weight generated according to the separability criterion (Equation (34)) was significant during the fraud classification process. The most relevant dimensions were successive outbidding, bidding ratio, and winning ratio (which makes sense when compared with the analysis in Figure 6). Figures 14 and 15 present a graphical evaluation based on a scatter plot of the two dimensions that best contribute to the class separability criterion.

Finally, the expansion of knowledge given by eFNN-SODA demonstrates a different way of interpreting the problem through fuzzy rules connected with the relevance of the problem's features. The Fuzzy Hoeffding Decision Tree model [66] also generated an interpretative approach, which presents a decision tree according to fuzzy techniques in constructing its leaves. The Expliclas [67] online solution (https://demos.citius.usc.es/ExpliClas/datasets (accessed on 12 August 2022)) facilitates the acquisition of knowledge to be compared with the extraction of knowledge obtained in this paper. The decision tree formed by the Fuzzy Hoeffding Decision Tree is shown in Figure 16.

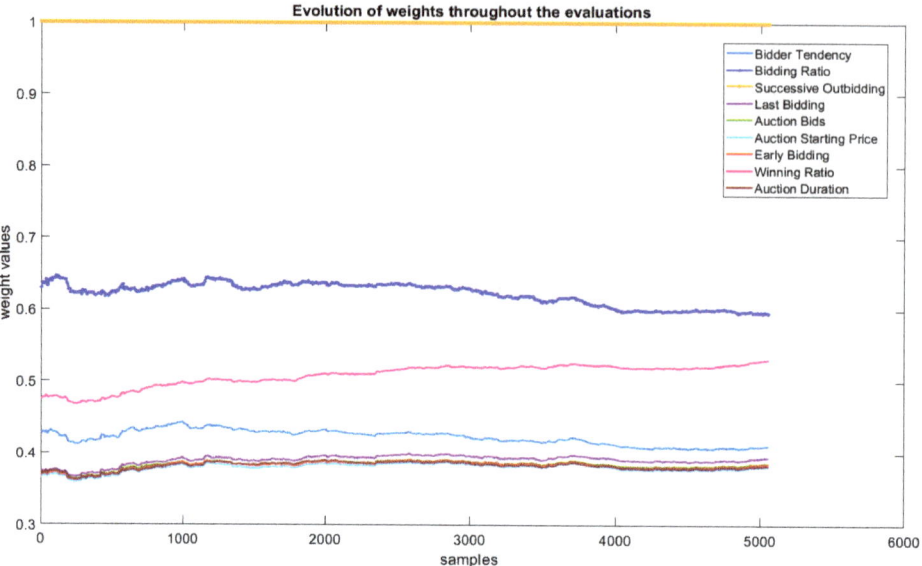

Figure 13. Feature separability criteria throughout the evaluations of the eFNN-SODA model.

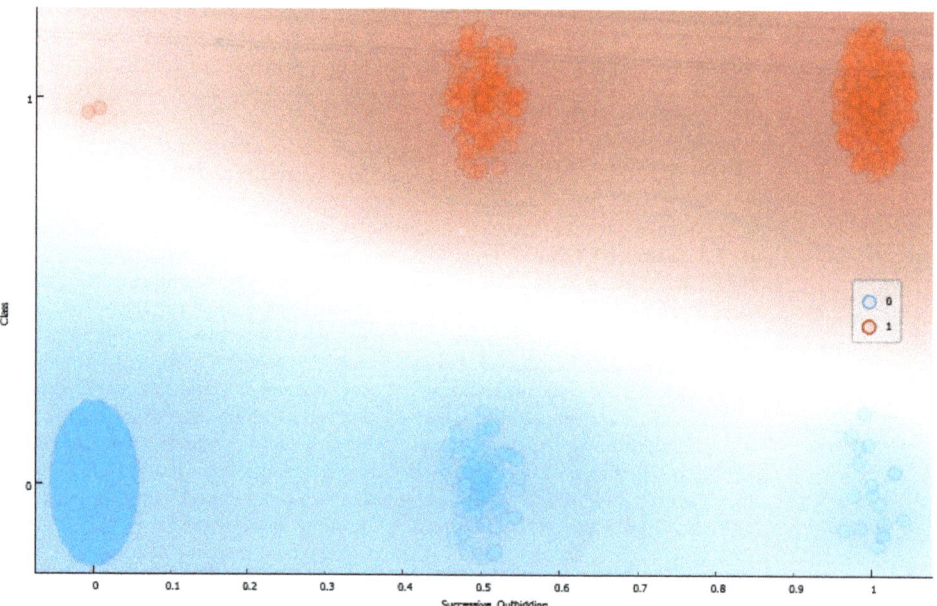

Figure 14. Scatter plot of successive outbidding feature.

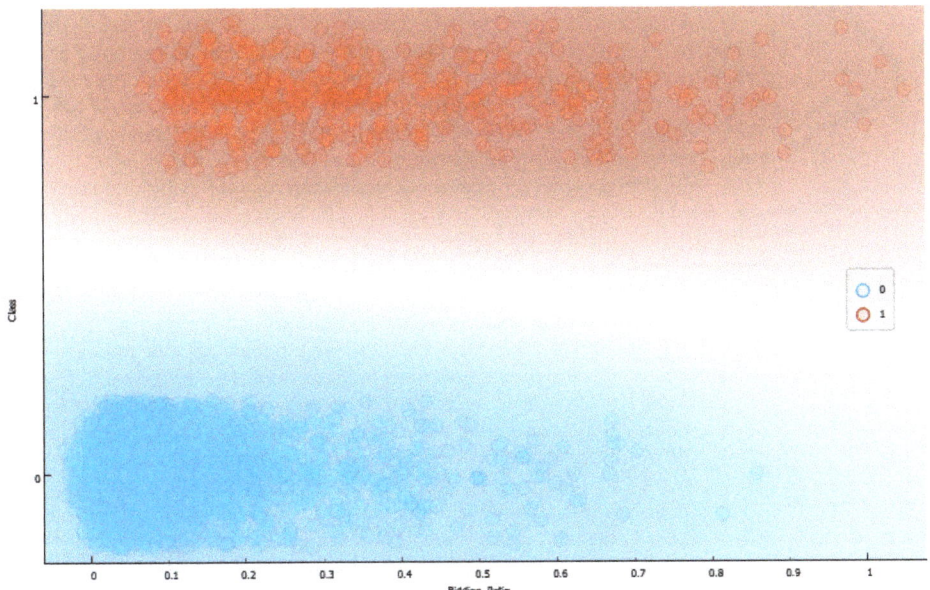

Figure 15. Scatter plot of bidding ratio feature.

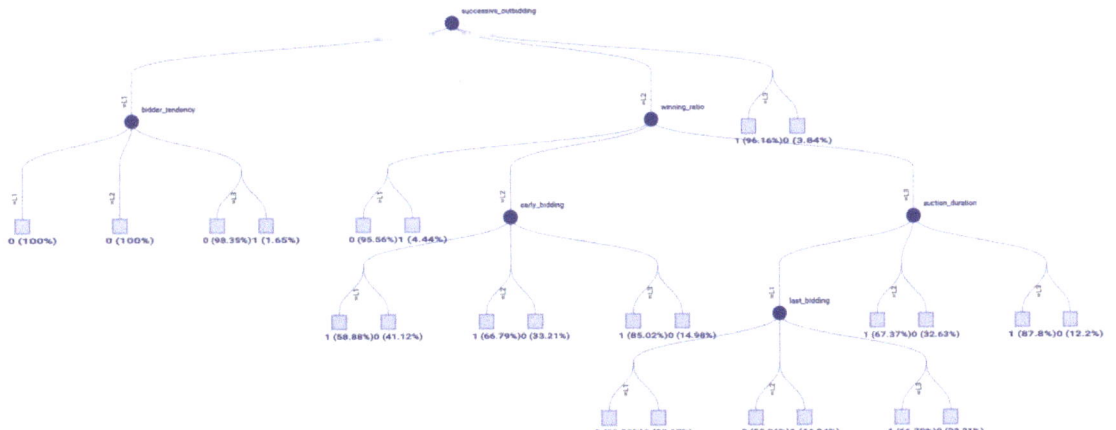

Figure 16. Fuzzy Hoeffding Decision Tree interpretability.

5. Discussions

Discussions about the results obtained by the model are carried out based on the interpretability criteria to be evaluated in eFNN-SODA. This is required to ensure that the information extraction stated below is trustworthy and coherent, in addition to verifying the accuracy findings.

The results obtained by the model present factors that can be interpreted through fuzzy rules that solve the target problem. The fuzzy rules were validated concerning the criteria of simplicity (fewer rules with greater accuracy), distinguishability (not similar), consistency, with coverage in all samples evaluated, and that meet the criteria of completeness. Through these generated fuzzy rules, it is possible to identify and analyze their antecedents and consequents and the relationship between them, and finally, it is possible to identify the evolution of the model through the creation of new fuzzy rules.

An interpretive evaluation of fuzzy rules leads us to identify that all dimensions of the problem (except successive outbidding) have similar final relevance values for separating fraudulent behavior from normal behavior. This indicates that there is not such a clear separability between the two groups of samples in these dimensions. The evolving behavior of the weights (Figure 13) also demonstrates stability in the order of relevance of the features of the problem. They did not change their position of relevance throughout the experiment. The second highest weight value for the feature weight technique is the Bidding Ratio of 0.59. This demonstrates that, except for the Successive Outbidding dimension, all the others involved in the problem have poorly separable samples according to the Dy–Brodley separability criteria.

The comparison of interpretability extracted by the eFNN-SODA model and the Fuzzy Hoeffding Decision Tree model (Figure 16) has some similarities and distinctions. The tree model is easier for visualizing the relationships, but the fuzzy rule model is closer to human reasoning. Regarding their similarities to the root of the tree shown in Figure 16 that has the Successive Outbidding as its central dimension, other nodes arise for the following evaluations. This corroborates that this is the first value to be analyzed in the tree, thus being the most relevant dimension. The generated fuzzy rules also identify this factor, as Successive Outbidding significantly impacts rule antecedents.

The three most relevant dimensions to the problem were determined based on the fuzzy rules generated. With them, it is possible to elaborate future studies and technological tools that can avoid Successive Outbidding, Bidding Ratio, and Winning Ratio considered fraudulent. Successive Outbidding Technologies to avoid subsequent bids can be implemented in digital solutions that provide auction services. For example, if the fuzzy neural network identifies a user at high risk of fraud, they cannot place successive bids for a certain period. In the same way, when possible fraud in the bids is identified, they may undergo an audit (human or through some expert system) or be the subject of extra validation, such as the confirmation of some data or limitation on the number of bids that they can give for a product until their veracity is proven. Finally, different mechanisms can be provided for Winning Ratio fraud if the fuzzy neural network identifies that that winner has a fraudulent profile.

6. Conclusions

This work presented an evaluation through evolving fuzzy neural network models to address auction fraud problems. The eFNN-SODA model proposed in this paper achieved more than 98% in correctly classifying fraudulent situations or not within a highly unbalanced data set. eFNN-SODA can be considered the most straightforward approach compared with the other models evaluated in the experiment, as it performed the fraud classification tasks with fewer fuzzy rules and greater assertiveness. The model also presented distinguishable fuzzy rules because their antecedents and consequents are dissimilar.

Another factor verified in the experiments is that the inconsistencies of the fuzzy rules during the experiment were minimal and promptly corrected by training eFNN-SODA. Regarding the coverage criterion, as they are Gaussian neurons formed in the fuzzification process, eFNN-SODA also meets the requirement. Finally, the model meets the criterion of completeness throughout the experiment. Therefore, it can be concluded that the model's fuzzy rules presented at the end of the training meet the criteria of consistency, coverage, and completeness.

The evaluation of problem features helped the classifier to correctly identify the classes by assigning to each fuzzy rule antecedent a weight corresponding to its relevance in correctly identifying the problem classes. Factors such as the expansion of knowledge through a new approach, the visualization of the interpretability of the problem, and the analysis of antecedents and consequents were also presented in this article. They corroborate the statement that the fuzzy rules generated by the eFNN-SODA model are efficient and interpretable in solving fraud identification problems in an auction.

The model proposed in this paper can generate technological products that evolve their knowledge as the behavior of fraud changes. This can happen by creating auxiliary expert systems to validate or investigate certain behaviors in the auctions. Fuzzy rules can be shown to system administrators and users, allowing corrective or analysis actions. With this type of system, both parties have clear and understandable logical relationships. Finally, creating an expert system to assist in data validation after the auction is also possible.

Possible future work can be conducted to develop new training techniques for model evolution and to use other neural structures to improve model accuracy. Another interesting topic to address is to evaluate or propose new methods for verifying the interpretability of models evolving neuro-fuzzy networks. The validation of the generated fuzzy rules can also be a potential focus of future research. Other (less unbalanced) data sets and auction fraud assessments are also encouraged to verify the model's adaptability in solving the problems of this behavior.

Author Contributions: Conceptualization, P.V.d.C.S. and E.L.; methodology, P.V.d.C.S. and H.R.B.; software, P.V.d.C.S.; validation, E.L.; formal analysis, E.L.; investigation, P.V.d.C.S. and A.J.G.; resources, P.V.d.C.S.; data curation, E.L.; writing—original draft preparation, P.V.d.C.S., H.R.B., and A.J.G.; writing—review and editing, E.L. and H.R.B.; visualization, P.V.d.C.S., A.J.G., and H.R.B.; supervision, E.L.; project administration, E.L.; funding acquisition, E.L. All authors have read and agreed to the published version of the manuscript.

Funding: Open Access Funding by the Austrian Science Fund (FWF), contract number P32272-N38, acronym IL-EFS.

Data Availability Statement: Publicly available datasets were analyzed in this study. These data can be found here: https://archive.ics.uci.edu/ml/datasets/Shill+Bidding+Dataset (accessed on 12 August 2022).

Conflicts of Interest: The authors declare no conflict of interest.

References

1. Mitchell, R.; Michalski, J.; Carbonell, T. *An Artificial Intelligence Approach*; Springer: Berlin/Heidelberg, Germany, 2013.
2. Chua, C.E.H.; Wareham, J. Fighting internet auction fraud: An assessment and proposal. *Computer* **2004**, *37*, 31–37. [CrossRef]
3. Button, M.; Nicholls, C.M.; Kerr, J.; Owen, R. Online frauds: Learning from victims why they fall for these scams. *Aust. N. Z. J. Criminol.* **2014**, *47*, 391–408. [CrossRef]
4. Alzahrani, A.; Sadaoui, S. Scraping and preprocessing commercial auction data for fraud classification. *arXiv* **2018**, arXiv:1806.00656.
5. Alzahrani, A.; Sadaoui, S. Clustering and labeling auction fraud data. In *Data Management, Analytics and Innovation*; Springer: Berlin/Heidelberg, Germany, 2020; pp. 269–283.
6. Buckley, J.; Hayashi, Y. Fuzzy neural networks: A survey. *Fuzzy Sets Syst.* **1994**, *66*, 1–13. [CrossRef]
7. De Campos Souza, P.V. Fuzzy neural networks and neuro-fuzzy networks: A review the main techniques and applications used in the literature. *Appl. Soft Comput.* **2020**, *92*, 106275. [CrossRef]
8. Novák, V. Fuzzy natural logic: Towards mathematical logic of human reasoning. In *Towards the Future of Fuzzy Logic*; Springer: Berlin/Heidelberg, Germany, 2015; pp. 137–165.
9. Gu, X.; Angelov, P.; Kangin, D.; Principe, J. Self-Organised direction aware data partitioning algorithm. *Inf. Sci.* **2018**, *423*, 80–95. [CrossRef]
10. Huang, G.B.; Zhu, Q.Y.; Siew, C.K. Extreme learning machine: Theory and applications. *Neurocomputing* **2006**, *70*, 489–501. [CrossRef]
11. De Campos Souza, P.V.; Lughofer, E. An evolving neuro-fuzzy system based on uni-nullneurons with advanced interpretability capabilities. *Neurocomputing* **2021**, *451*, 231–251. [CrossRef]
12. De Campos Souza, P.V.; Lughofer, E.; Guimaraes, A.J. Evolving Fuzzy Neural Network Based on Uni-nullneuron to Identify Auction Fraud. In Proceedings of the Joint Proceedings of the 19th World Congress of the International Fuzzy Systems Association (IFSA), the 12th Conference of the European Society for Fuzzy Logic and Technology (EUSFLAT), and the 11th International Summer School on Aggregation Operators (AGOP), Bratislava, Slovakia, 19–24 September 2021; pp. 314–321. [CrossRef]
13. Haykin, S.S.; Haykin. *Neural Networks and Learning Machines*; Pearson Education Upper Saddle River: Hoboken, NJ, USA, 2009; Volume 3.
14. Fausett, L.V. *Fundamentals of Neural Networks: Architectures, Algorithms, and Applications*; Prentice-Hall Englewood Cliffs: Hoboken, NJ, USA, 1994; Volume 3.
15. Zhang, Z. Artificial neural network. In *Multivariate Time Series Analysis in Climate and Environmental Research*; Springer: Berlin/Heidelberg, Germany, 2018; pp. 1–35.

16. Rumelhart, D.E.; Durbin, R.; Golden, R.; Chauvin, Y. Backpropagation: The basic theory. *Backpropagation Theory Archit. Appl.* **1995**, 1–34.
17. Pedrycz, W. Neurocomputations in relational systems. *IEEE Trans. Pattern Anal. Mach. Intell.* **1991**, *13*, 289–297. [CrossRef]
18. Mithra, K.; Sam Emmanuel, W. GFNN: Gaussian-Fuzzy-Neural network for diagnosis of tuberculosis using sputum smear microscopic images. *J. King Saud Univ.-Comput. Inf. Sci.* **2021**, *33*, 1084–1095. [CrossRef]
19. Hu, Q.; Gois, F.N.B.; Costa, R.; Zhang, L.; Yin, L.; Magaia, N.; de Albuquerque, V.H.C. Explainable artificial intelligence-based edge fuzzy images for COVID-19 detection and identification. *Appl. Soft Comput.* **2022**, *123*, 108966. [CrossRef]
20. Salimi-Badr, A.; Ebadzadeh, M.M. A novel learning algorithm based on computing the rules' desired outputs of a TSK fuzzy neural network with non-separable fuzzy rules. *Neurocomputing* **2022**, *470*, 139–153. [CrossRef]
21. Nasiri, H.; Ebadzadeh, M.M. MFRFNN: Multi-Functional Recurrent Fuzzy Neural Network for Chaotic Time Series Prediction. *Neurocomputing* **2022**, *507*, 292–310. [CrossRef]
22. Pan, Q.; Li, X.; Fei, J. Adaptive Fuzzy Neural Network Harmonic Control with a Super-Twisting Sliding Mode Approach. *Mathematics* **2022**, *10*, 1063. [CrossRef]
23. Amirkhani, A.; Shirzadeh, M.; Molaie, M. An Indirect Type-2 Fuzzy Neural Network Optimized by the Grasshopper Algorithm for Vehicle ABS Controller. *IEEE Access* **2022**, *10*, 58736–58751. [CrossRef]
24. Algehyne, E.A.; Jibril, M.L.; Algehainy, N.A.; Alamri, O.A.; Alzahrani, A.K. Fuzzy Neural Network Expert System with an Improved Gini Index Random Forest-Based Feature Importance Measure Algorithm for Early Diagnosis of Breast Cancer in Saudi Arabia. *Big Data Cogn. Comput.* **2022**, *6*, 13. [CrossRef]
25. Novák, V.; Perfilieva, I.; Dvorak, A. *Insight into Fuzzy Modeling*; John Wiley & Sons: Hoboken, NJ, USA, 2016.
26. Murinová, P.; Novák, V. The theory of intermediate quantifiers in fuzzy natural logic revisited and the model of "Many". *Fuzzy Sets Syst.* **2020**, *388*, 56–89. [CrossRef]
27. Novák, V.; Perfilieva, I. Forecasting direction of trend of a group of analogous time series using F-transform and fuzzy natural logic. *Int. J. Comput. Intell. Syst.* **2015**, *8*, 15–28. [CrossRef]
28. Nguyen, L. Integrating The Probabilistic Uncertainty to Fuzzy Systems in Fuzzy Natural logic. In Proceedings of the 2020 12th International Conference on Knowledge and Systems Engineering (KSE), Can Tho City, Vietnam, 12–14 November 2020; pp. 142–146.
29. Xu, X.; Ding, X.; Qin, Z.; Liu, Y. Classification Models for Medical Data with Interpretative Rules. In *Proceedings of the International Conference on Neural Information Processing*; Springer: Berlin/Heidelberg, Germany, 2021; pp. 227–239.
30. Nemat, R. Taking a look at different types of e-commerce. *World Appl. Program.* **2011**, *1*, 100–104.
31. Trevathan, J. Getting into the mind of an "in-auction" fraud perpetrator. *Comput. Sci. Rev.* **2018**, *27*, 1–15. [CrossRef]
32. Abidi, W.U.H.; Daoud, M.S.; Ihnaini, B.; Khan, M.A.; Alyas, T.; Fatima, A.; Ahmad, M. Real-Time Shill Bidding Fraud Detection Empowered With Fussed Machine Learning. *IEEE Access* **2021**, *9*, 113612–113621. [CrossRef]
33. Anowar, F.; Sadaoui, S. Detection of Auction Fraud in Commercial Sites. *J. Theor. Appl. Electron. Commer. Res.* **2020**, *15*, 81–98. [CrossRef]
34. Anowar, F.; Sadaoui, S.; Mouhoub, M. Auction Fraud Classification Based on Clustering and Sampling Techniques. In Proceedings of the 2018 17th IEEE International Conference on Machine Learning and Applications (ICMLA), Orlando, FL, USA, 17–20 December 2018; pp. 366–371.
35. De Campos Souza, P.V.; Lughofer, E. Evolving fuzzy neural classifier that integrates uncertainty from human-expert feedback. *Evol. Syst.* **2022**, 1–23. [CrossRef]
36. Škrjanc, I.; Iglesias, J.A.; Sanchis, A.; Leite, D.; Lughofer, E.; Gomide, F. Evolving fuzzy and neuro-fuzzy approaches in clustering, regression, identification, and classification: A survey. *Inf. Sci.* **2019**, *490*, 344–368. [CrossRef]
37. Lughofer, E. On-line assurance of interpretability criteria in evolving fuzzy systems—Achievements, new concepts and open issues. *Inf. Sci.* **2013**, *251*, 22–46. [CrossRef]
38. De Campos Souza, P.V.; Lughofer, E. An advanced interpretable Fuzzy Neural Network model based on uni-nullneuron constructed from n-uninorms. *Fuzzy Sets Syst.* **2020**, *426*, 1–26. [CrossRef]
39. Lughofer, E. On-line incremental feature weighting in evolving fuzzy classifiers. *Fuzzy Sets Syst.* **2011**, *163*, 1–23. [CrossRef]
40. Angelov, P.; Gu, X.; Príncipe, J. A generalized methodology for data analysis. *IEEE Trans. Cybern.* **2017**, *48*, 2981–2993. [CrossRef]
41. Angelov, P.; Gu, X.; Kangin, D. Empirical data analytics. *Int. J. Intell. Syst.* **2017**, *32*, 1261–1284. [CrossRef]
42. Gu, X.; Angelov, P.P. Self-organising fuzzy logic classifier. *Inf. Sci.* **2018**, *447*, 36–51. [CrossRef]
43. Gu, X.; Angelov, P.P.; Príncipe, J.C. A method for autonomous data partitioning. *Inf. Sci.* **2018**, *460*, 65–82. [CrossRef]
44. Watson, D.F. Computing the n-dimensional Delaunay tessellation with application to Voronoi polytopes. *Comput. J.* **1981**, *24*, 167–172. [CrossRef]
45. Angelov, P.; Yager, R. A new type of simplified fuzzy rule-based system. *Int. J. Gen. Syst.* **2012**, *41*, 163–185. [CrossRef]
46. De Campos Souza, P.V.; Lughofer, E.; Rodrigues Batista, H. An Explainable Evolving Fuzzy Neural Network to Predict the k Barriers for Intrusion Detection Using a Wireless Sensor Network. *Sensors* **2022**, *22*, 5446. [CrossRef]
47. Dy, J.; Brodley, C. Feature Selection for Unsupervised Learning. *J. Mach. Learn. Res.* **2004**, *5*, 845–889.
48. Qin, S.; Li, W.; Yue, H. Recursive PCA for Adaptive Process Monitoring. *J. Process Control* **2000**, *10*, 471–486.
49. Klement, E.P.; Mesiar, R.; Pap, E. *Triangular Norms*; Springer Science & Business Media: Berlin, Germany, 2013; Volume 8.
50. Hirota, K.; Pedrycz, W. OR/AND neuron in modeling fuzzy set connectives. *IEEE Trans. Fuzzy Syst.* **1994**, *2*, 151–161. [CrossRef]

51. Pedrycz, W.; Reformat, M.; Li, K. OR/AND neurons and the development of interpretable logic models. *IEEE Trans. Neural Netw.* **2006**, *17*, 636–658. [CrossRef]
52. Huang, G.B.; Chen, L.; Siew, C.K. Universal approximation using incremental constructive feedforward networks with random hidden nodes. *IEEE Trans. Neural Netw.* **2006**, *17*, 879–892. [CrossRef]
53. Albert, A. *Regression and the Moore-Penrose Pseudoinverse*; Elsevier: Amsterdam, The Netherlands, 1972.
54. Rosa, R.; Gomide, F.; Dovzan, D.; Skrjanc, I. Evolving neural network with extreme learning for system modeling. In Proceedings of the 2014 IEEE Conference on Evolving and Adaptive Intelligent Systems (EAIS), Linz, Austria, 2–4 June 2014; pp. 1–7.
55. Lughofer, E. *Evolving Fuzzy Systems—Methodologies, Advanced Concepts and Applications*; Springer: Berlin/Heidelberg, Germany, 2011.
56. Lughofer, E.; Bouchot, J.L.; Shaker, A. On-line elimination of local redundancies in evolving fuzzy systems. *Evol. Syst.* **2011**, *2*, 165–187. [CrossRef]
57. Demšar, J.; Leban, G.; Zupan, B. FreeViz—An intelligent multivariate visualization approach to explorative analysis of biomedical data. *J. Biomed. Inform.* **2007**, *40*, 661–671. [CrossRef]
58. Hoffman, P.; Grinstein, G.; Marx, K.; Grosse, I.; Stanley, E. DNA visual and analytic data mining. In Proceedings of the Visualization '97 (Cat. No. 97CB36155), Phoenix, AZ, USA, 19–24 October 1997; pp. 437–441. [CrossRef]
59. de Campos Souza, P.V.; Torres, L.C.B.; Guimaraes, A.J.; Araujo, V.S.; Araujo, V.J.S.; Rezende, T.S. Data density-based clustering for regularized fuzzy neural networks based on nullneurons and robust activation function. *Soft Comput.* **2019**, *23*, 12475–12489. [CrossRef]
60. Souza, P.; Ponce, H.; Lughofer, E. Evolving fuzzy neural hydrocarbon networks: A model based on organic compounds. *Knowl.-Based Syst.* **2020**, *203*, 106099. [CrossRef]
61. De Campos Souza, P.V.; Rezende, T.S.; Guimaraes, A.J.; Araujo, V.S.; Batista, L.O.; da Silva, G.A.; Silva Araujo, V.J. Evolving fuzzy neural networks to aid in the construction of systems specialists in cyber attacks. *J. Intell. Fuzzy Syst.* **2019**, *36*, 6743–6763. [CrossRef]
62. Soares, E.; Angelov, P.; Gu, X. Autonomous Learning Multiple-Model zero-order classifier for heart sound classification. *Appl. Soft Comput.* **2020**, *94*, 106449. [CrossRef]
63. Angelov, P.; Gu, X. Autonomous learning multi-model classifier of 0-Order (ALMMo-0). In Proceedings of the 2017 Evolving and Adaptive Intelligent Systems (EAIS), Ljubljana, Slovenia, 31 May–2 June 2017; pp. 1–7. [CrossRef]
64. Leite, D.; Costa, P.; Gomide, F. Evolving granular neural network for semi-supervised data stream classification. In Proceedings of the 2010 International Joint Conference on Neural Networks (IJCNN), Barcelona, Spain, 18–23 July 2010; pp. 1–8. [CrossRef]
65. Bifet, A.; Holmes, G.; Kirkby, R.; Pfahringer, B. MOA: Massive Online Analysis. *J. Mach. Learn. Res.* **2010**, *11*, 1601–1604.
66. Ducange, P.; Marcelloni, F.; Pecori, R. Fuzzy Hoeffding Decision Tree for Data Stream Classification. *Int. J. Comput. Intell. Syst.* **2021**, *14*, 946–964. [CrossRef]
67. Alonso, J.M.; Bugarín, A. ExpliClas: Automatic Generation of Explanations in Natural Language for Weka Classifiers. In Proceedings of the 2019 IEEE International Conference on Fuzzy Systems (FUZZ-IEEE), Hyderabad, India, 7–10 July 2019; pp. 1–6. [CrossRef]

MDPI
St. Alban-Anlage 66
4052 Basel
Switzerland
Tel. +41 61 683 77 34
Fax +41 61 302 89 18
www.mdpi.com

Mathematics Editorial Office
E-mail: mathematics@mdpi.com
www.mdpi.com/journal/mathematics

www.ingramcontent.com/pod-product-compliance
Lightning Source LLC
LaVergne TN
LVHW070609100526
838202LV00012B/600